Stephen Alfred Forbes

Experiments with the muscardine Disease of the Chinch-Bug

And with the Trap and Barrier Method for the Destruction of that Insect

Stephen Alfred Forbes

Experiments with the muscardine Disease of the Chinch-Bug
And with the Trap and Barrier Method for the Destruction of that Insect

ISBN/EAN: 9783337062552

Printed in Europe, USA, Canada, Australia, Japan

Cover: Foto ©ninafisch / pixelio.de

More available books at **www.hansebooks.com**

UNIVERSITY OF ILLINOIS.

Agricultural Experiment Station.

. URBANA, MARCH, 1895.

BULLETIN NO. 38.

EXPERIMENTS WITH THE MUSCARDINE DISEASE OF THE CHINCH-BUG, AND WITH THE TRAP AND BARRIER METHOD FOR THE DESTRUCTION OF THAT INSECT.

The history of chinch-bug injury in Illinois is substantially that of a succession of waves of increase which slowly rise to a highest point and then rapidly fall away to insignificance, the rise of the wave usually occupying from three to five years or more, and its recession commonly requiring only one or two. The year 1894 was marked by the culmination of such a period of increase, the wave of destruction reaching very nearly its highest recorded point, and covering a large part of this state from the Ohio River to the northern tier of counties. It also extended beyond our borders into Missouri, Kansas, and southeastern Iowa. In Illinois it was most injurious (1) in the southern and south-central part of the state, (2) in the western-central counties, and (3) in a few counties near the northern boundary—being practically harmless or nearly so only in the eastern part of central and north-central Illinois.

Complaints of serious injury and appeals for aid were received at this office during the year from six hundred and ten* towns in seventy-six counties—a number not previously equaled since 1887. The greater part of these appeals took the form of applications for material with which to introduce the contagious insect diseases into infested fields. While the results of our previous experimental work with the principal fungous disease of the chinch-bug were not favorable to the

---

*In Bulletin 5 from the State Entomologist's office this number was incorrectly given as five hundred.

idea that it would be found to have any considerable value as a means of arresting injury by the chinch-bug where conditions were particularly favorable to the multiplication and maintenance of that insect, I was nevertheless induced to undertake to supply this demand, largely by the following considerations:

1. Notwithstanding our previous experience, I was not yet prepared to say positively that the contagious-disease method if persistently followed up would not take effect in very many cases even under ordinary circumstances; and as long as there was even an appreciable chance that the farmers might thus save any considerable part of their crops this season by our aid it seemed to me that they were entitled to the benefit of the doubt in favor of this procedure, especially as the expense of a general distribution would be at most a trifle compared with the great interests at stake.

2. The general credit which this method has received through the agricultural papers and the daily press, as well as through several state official publications, and the firm belief which very many of our farmers already had in it, made it seem very likely that nothing would satisfy them except a chance to try it.

3. I was fairly well assured, as a result of our own field observations and laboratory experiments, that under favorable weather conditions this contagious disease might do an immense service to those parts of the state threatened with the destruction of wheat and corn; and as we could not foresee the weather of the season, I thought it incumbent on me to take measures to derive the greatest possible advantage from weather favorable to the spread of the disease, if such weather should follow.

4. I wished, finally, to see for myself how generally and accurately the somewhat complicated directions necessary to an intelligent use of this method would be followed out by the average farmer when greatly interested in the result.

The demand for contagion material became so great by June 1 that it was evident that I should no longer be able to meet it from current appropriations at my disposal and with the aid of my usual corps of assistants. I consequently suggested, early in June, to the authorities of the State Agricultural Experiment Station, at Urbana, the idea of providing for more elaborate experiments in the field, and of supplying a limited amount of tested infection material for trial by farmers themselves. This plan of experimentation and distribution was very promptly taken up and favorably considered by the executive committee of the Station board, and I was authorized to spend in this direction not to exceed $200 previous to the meeting of the board June 1, and subsequent to that meeting $250 more. I consequently engaged the necessary assistants, enlarged our facilities, and published a general notice to those interested of my willingness to receive live chinch-bugs and to return infected ones in their place. I used for this purpose the Associated Press, June 5, 1894, and also sent out June 7, through the Experiment Station office, a press bulletin on "The Chinch-bug in Illinois." This offer was most eagerly accepted by a very large number of farmers, and we were presently very nearly overwhelmed—as were also the local express offices and the post office—by packages of chinch-bugs arriving from all parts of the state, and in all imaginable conditions.

In order to avail myself of the much larger experience of the Kansas University Station, I followed precisely at first the infection methods there in use, depending upon an exposure of the chinch-bugs to insects dead with the disease and covered with the characteristic fungous growth; and to make assurance doubly sure I had already obtained a supply of material directly from the Kansas State University, although we had the same fungus in our own infection boxes at the time. Notwithstanding the great enlargement of our facilities, and the continuous expert attention which the whole subject received, especially from Mr. John Marten, who has had principal charge of our disease experiments for four years, the contagion did not spread rapidly enough in our boxes to make it possible to meet at once more than a small percentage of the demand. I found later that a part of this slow development was due to a difficulty which seems not to have been previously noticed by any one here or elsewhere; namely, the appearance in our contagion boxes of swarms of minute mites which fed upon the fungus as fast as it was developed.

Next, observing that the thirteen-year locusts (*Cicada tredecim*), a brood of which was rapidly disappearing, had many of them died with this disease, and bore a profuse growth of the characteristic fungus in excellent condition, I had a large quantity of these collected, and used these dead locusts for distribution, accompanied in each case by chinch-bugs which had been exposed to the infection.

Finally, having ascertained, as a result of experiments made previously and also at the time, that the 'cultivated fungus grown upon a mixture of corn meal and beef broth was apparently as effective for the destruction of chinch-bugs as that obtained from the insects themselves, I had a large quantity grown artificially on this material, and used this also for distribution.

By these methods I succeeded, by about the 20th of July, in supplying all who had sent requests up to the 10th of that month—a little over two thousand for the season. As I had issued a second bulletin June 30, giving notice that it would be impossible to continue the distribution beyond July 10, I considered the obligations I had assumed thus fulfilled, and this work was brought practically to an end.

Each lot of chinch-bugs, living and dead, was accompanied by the following circular of directions for their utilization, and of caution against hasty observation and inference:

"DEAR SIR: I send you by this mail chinch-bugs which have been successfully exposed to the white fungus disease of that insect, and are in a condition to convey it to others.

"To propagate this disease in your field, make a tight shallow wooden box, say, 24x36x6 inches, and place in it a layer of dirt half an inch deep, free from leaves or other rubbish. Moisten this dirt without making it muddy, and then put in a thin layer of green wheat or corn. Scatter the dead chinch-bugs sent you over the bottom of the box, and shut up with them a quantity of live bugs from the field—as many as can well move about in the box without being anywhere more than one layer deep. Fasten the cover down tight, so that nothing can escape, and set the box where it will be protected from sun and wind. A cellar or a basement room is to be preferred.

"Open the box daily and moisten its sides and contents (without making them muddy) when they begin to get dry, and also change the food as that in the box becomes yellow. When it is seen that the white, mouldy bugs are becoming more numerous, probably in about three or four days, take a part of the bugs, dead and alive, out of the box, putting in fresh live ones to take their places, and close the box as before.

"Those taken out should then be scattered through the infested field where the bugs are thickest—at the bases of the leaves in the corn fields, around the lower ends of the stalks, and the like. Make this distribution, by preference, in the evening, when the dew is on, or, still better, just after a rain, and repeat if dry weather follows. Continue these collections and distributions as above through the whole season, making certain each time chinch-bugs are taken out that white ones are left in the box; and when winter comes put all the dead bugs remaining into pill boxes for use the following year.

"Those wishing to form an independent judgment of the practical value of this method of dealing with chinch-bugs should take into account the following facts:

"1. The white fungus causing insect disease requires moisture for its full development, and especially for the formation of the minute "spores" by whose dispersal the disease is conveyed from one insect to another. In times of severe drought it propagates slowly or not at all.

"2. It takes effect on a weakened insect more readily than on one in full vigor; on the full grown chinch-bug more easily than on the young; and hence most easily of all on spent adults which have already laid their eggs and are about to perish by the natural termination of their life period.

"3. It is a native disease of the chinch-bug and never dies out entirely, but is likely to appear spontaneously over a large extent of country when conditions favorable to its development are long maintained.

"4. Two generations of the chinch-bug appear each year, and when each of these generations matures, the adult bugs commonly take wing and scatter, thus disappearing largely from fields or parts of fields heavily infested by them. Such dispersal has often been mistaken for a destruction of chinch-bugs by disease. One generation matures shortly after wheat harvest and the other in late summer and in the fall.

"5. The chinch-bug sheds its skin four times while growing, and the empty skins left by it are often mistaken for dead bugs—a mistake which has sometimes led to a false conclusion as to the effect of these infection experiments. The cast skins never bear wings, as the insect does not moult after its wings are formed. They may further be readily distinguished from the dead bugs by the fact that when pressed between the thumb-nails they are readily seen to be empty shells without contents.

"To judge intelligently of the effect of any attempt to introduce disease, the observer should examine very carefully, in advance, the field in which the experiment is to be tried, and adjacent fields as well, to see whether bugs dead with the white fungus may not already be present. If the disease appears at the point where the infected chinch-bugs are placed, he should repeat this general examination, and make sure that the disease may not have occurred spontaneously and without special reference to his experimental introduction of it. He should also notice whether young bugs (those without wings) are attacked by it, as, if they are not, it is quite likely it is only carrying away those about to die of old age. On the other hand, it should be remembered that these especially susceptible adult bugs may afford the best means of securing a general dissemination of the fungus in the fields, where it may lie dormant for a considerable time, ready to spring into sudden activity when favorable weather conditions appear.

"Advantage should be taken of every considerable shower, and especially of every long rain, to scatter the diseased bugs, and all fields under observation should be thoroughly inspected some two or three days thereafter."

I was also careful in every published statement or written communication on the subject to warn all against reliance upon this method to the neglect of other preventive or destructive measures, and emphasized in every way its purely experimental character.

In the meantime, experiments, carefully planned and closely followed up, were made in the field through Mr. Marten and Mr. Johnson, both assistants of the office, by the distribution, in wheat and corn fields, of fungus cultures and of chinch-bugs dead with disease and bearing the characteristic fungus in a fruiting condition. One series of such experiments was made on the University Experiment Station farm, at Urbana, and others were set on foot at several points in southern Illinois, each being followed up by repeated visits made to ascertain the result.

The opportunity was improved during these visits to examine also several experiments made by farmers of our acquaintance with material obtained from the office under such conditions and management as to give them positive value.

Pressing and engrossing as was the series of field operations undertaken this year, our experimental work was not confined to these, but laboratory experiments directed to special ends were carried on during some weeks by the aid of Miss Nettie Ayers, a recent assistant in the bacteriological laboratory of the University of Illinois. Artificial cultures, under varying conditions, of the Sporotrichum characteristic of the chinch-bug disease, and of another species of parasitic fungus obtained from a correspondent, were made by Miss Ayers, and the results of such cultures were tested upon chinch-bugs, living and dead, upon cabbage worms both living and dead, and upon a variety of other insects, these experiments being so managed that the conditions under which they were made were precisely subject to our control.

In addition to these various experiments with the contagion method, I made at Urbana this summer a thorough test of certain measures for the arrest and destruction of chinch-bugs as they moved from wheat to corn in early June and July.

The entire series, for 1894, of these experimental studies in the laboratory and in the field are here reported, whether made under the immediate auspices of the Agricultural Experiment Station, or as a part of the regular work of the State Laboratory of Natural History. The detailed description of experiments here given is preceded by an outline of their arrangement and subordination, showing in every case their connection with each other and the entire history of the material used in each experiment, and also by a classified list of the experiments, in which they are grouped according to their objects.

I now propose to state and discuss the general conclusions to be drawn from this work, with such references to individual experiments as will enable the critical reader to judge of the character and weight of the evidence upon which these conclusions rest.

GENERAL DISCUSSION OF RESULTS.

The more important results of the season's experiments which have an economic value may be briefly summarized in the following terms:

1. The white muscardine will not spread among vigorous chinch-bugs in the field in very dry weather to an extent to give this disease any practical value as a means of promptly arresting serious chinch-bug injury under such conditions. (See Nos. 55 to 58, and 61, 62, 77, etc.) On the contrary, even when it has appeared spontaneously, or as a result of artificial measures for its introduction, it may be completely arrested by dry weather, remaining in abeyance at least until the weather changes. (See No. 53, June 5 and June 20; No. 55, June 7, June 19, and August 8; No. 57, concluding discussion; Nos. 60, 63–67, etc.)

2. It is most likely to "catch" in low spots, where the soil is kept somewhat moist by dense vegetation, a mat of fallen herbage, or the like. Shocks of corn, especially when the crop is cut early, furnish excellent places for the development of this disease. (See No. 55, June 20; No. 57, September 18, 19, and 28; and Nos. 76 and 77.) Indeed, the presence in any field, of spots especially favorable to the growth of the Sporotrichum infection seemed, according to our observations, to have much more to do with the appearance and spread of the white muscardine among chinch-bugs than even the most persistent distribution of dead or infected specimens in the absence of such natural culture beds—a fact which contains the suggestion of a new method for the propagation and dissemination of this disease. It will be well worth while, consequently, to try the effect of excessive moisture and an inviting shelter on here and there a spot in an infested field, such as might be afforded by an overgrowth of small grain produced by heavy fertilization; or by trampling down a few hills of corn; or by the early cutting and shocking of some small part of the crop. If no spontaneous

development of muscardine were to follow, such spots would at any rate be excellent places to start a field infection.

3. If decidedly wet weather follows upon its introduction, even after an interval of several weeks, it is likely to start up and take visible effect; but continuous rains, depressing the vital energies of the insect, seem commonly requisite to its efficient action. (See Nos. 55 to 58, 77, etc.)

4. It is always so generally prevalent, in a more or less obscure condition, among chinch-bugs or other insects in Illinois, both north and south, that it is very likely to appear and spread, as if spontaneously, whenever conditions favorable to its development long prevail, whether it has been purposely introduced or not. (See especially No. 76.)

5. The time elapsing between the establishment of such favorable conditions and the full development of the disease among the chinch-bugs of any locality, may possibly be shortened if the infection has previously been introduced by artificial means; but our own experiments, it must be confessed, do not lend any material support to this supposition. (See No. 57, concluding discussion.)

6. Whatever weakens the insect favors its spread, as a rule. It is consequently much more likely to attack adults than young, especially spent males, and females which have laid their eggs, and which are soon to die of old age; but it nevertheless often kills young of all ages. From the record of our large contagion boxes (Nos. 68–71) it appears that after the establishment in my laboratory, July 1, of a special reception box into which all insects sent in by mail or express were put as received, the development of the fungus in the contagion boxes was much less rapid than before. The reception box was so managed that not only were all dead bugs excluded from the contagion boxes, but only the more vigorous of those remaining alive at the time of their arrival were transferred.

The supposed weakening effect of close confinement in a saturated atmosphere was also avoided in this reception box by leaving it open, the escape of the bugs being prevented by heavily chalking the inside of the box for four or five inches downward from the top. This chalk-band was renewed occasionally, as it was worn away by the chinch-bugs in their efforts to escape. The same device was used to confine the bugs in the contagion boxes when these were opened. The apparent effect of this elimination of weakened insects was greatly to diminish the number which succumbed to the muscardine infection.

In agreement with the above, we have noticed that the fall generation of adults is less subject to it, other things being equal, than the generation which matures in midsummer. As this fall brood is to live over winter before laying its eggs, it contains no worn out adults.

7. The fungus producing this disease will start rarely, if at all, on dead chinch-bugs, if we may judge from the results of several experiments made this summer (see Nos. 36–40 and Nos. 46 and 47). Where-

ever a dead chinch-bug shows the fungus in the field, it is therefore probable that it was infected while alive. Some doubt is thrown upon this conclusion, however, by the fact that upon dead soft-bodied insects, like cabbage worms, the Sporotrichum grew as promptly and luxuriantly as upon the insects infected while still alive. (See Nos. 21–27 and 41.)

8. The resistant power of healthy chinch-bugs exposed to infection is well shown by the fact that thousands of bugs, young and old, have commonly lived for many days, and even for several weeks, moulting, maturing, copulating, and laying their eggs, when shut up in contagion boxes which had been heavily stocked with fungus spores from dead insects and had been made in every way as favorable as possible to the development of the disease. The percentage of those that would succumb from day to day was often ridiculously small. (See Nos. 68–71.) On the other hand, it is probable that the heavy pressure upon the office for a supply of infected chinch-bugs frequently induced the too early and complete removal of the bugs from such boxes, thus retarding the development of the fungus among the imprisoned insects.

9. The growth of the fungus in such boxes is sometimes checked and the whole experiment brought to a standstill by the appearance in the boxes of minute mites (apparently brought in with the food supplied to the bugs), which multiply in the boxes and greedily devour the fungus of white muscardine as fast as it grows. (See No. 68, July 31, August 9 and 22; No. 69, July 30 and August 3; No. 70, July 30 and 31, and August 2; and No. 71, July 30.)

These mites were repeatedly noticed by us in July, but were not suspected of an injurious influence on our operations until July 30, when experiments made showed us that they were diligently feeding on the growing Sporotrichum. Confined with a fungus-covered chinch-bug July 30 at 3 P. M., they had completely cleared it off by the next morning. Another lot, placed under a glass with four such bugs at 9 A. M., had eaten up the last vestige of the fungus by 4:30 P. M. Similar trials showed that they would clear away with equal readiness the fungus growth from a culture on corn meal batter. Prolonged search of the earth outside, made where the supply for our contagion boxes was obtained, and a similar search of the sources of the food supply of the imprisoned chinch-bugs, gave us no hint of the origin of the mites. The same mite species was noticed August 7 in the contagion box of a farmer near Tonti, in southern Illinois, and it seems likely that these mites came in with the chinch-bugs sent us from the field.

10. Comparative experiments with fungus spores from diseased chinch-bugs and with those derived from artificial cultures on corn meal moistened with beef broth, show that the latter are nearly, if not quite, as efficient agents of infection as the former. We have used only culti-

vated spores two or three removes from the growth on the insect, and consequently are not prepared to say that continued cultivation on an inanimate medium might not finally diminish the virulence of the fungus parasite; but, on the other hand, we have no very good reason to suppose that this will prove to be the case; and I have no doubt that by a properly guarded procedure, these artificial cultures, which can easily be made in almost unlimited quantity, may be utilized for a dissemination of the spores of these insect diseases, with great advantage in convenience, expedition, and economy of operation. (Compare Nos. 3, 4, and 5 with No. 54; also No. 71 with No. 70, up to July 6.)

Comparative infection experiments with acid and neutral cultures were indeterminate in result, with the probability favoring the greater efficiency of the neutral cultures. (See Nos. 15–20, 28, and 51.)

11. The history of Experiment No. 1 and its derivatives shows beyond question the possibility of doing excellent work on chinch-bugs with fungus of this disease derived from other insect species. It is probable that many cases of its apparently spontaneous appearance among chinch-bugs are to be traced to such sources of infection. It was upon this ground that fragments of thirteen-year locusts profusely covered with Sporotrichum were distributed this summer, together with chinch-bugs previously exposed to infection, for experimental use by farmers.

12. A comparison of the infection experiments made on chinch-bugs with those made on cabbage worms shows clearly the very much greater susceptibility of the latter to Sporotrichum attack—a fact due possibly to their thinner skin and more juicy substance. Living and dead cabbage worms were infected with equal readiness if the air was kept moist. The spores started quickly on any part of the body, the growing hyphae penetrating the skin in one place seemingly as freely as in another. An external development of the fungus commonly became noticeable on the second day, as in artificial cultures. Cabbage worms were frequently, but not invariably, turned a dull red color by the growth of the Sporotrichum. In one experiment, which differed from the others by the omission of the layer of moist sand on the bottom of the dish in which the larvae were confined, this raspberry color was the only external evidence of successful infection with the Sporotrichum, no external growth appearing—a fact probably to be attributed to the comparative dryness of the air.

13. This is the place to make mention of certain experiments with the infection of insects in the laboratory which resulted in unusual developments of *Sporotrichum globuliferum*, illustrated by figures accompanying this report (Plate VII). With the exception of the two growths from June beetles (Fig. 5 and 6), whose botanical characters are identical with those of Sporotrichum, these figures were made from growths resulting from the infection of living insects with spores from cultures

made by us. The identity of these Isaria forms was further verified by raising the common Botrytis form from them on agar.

In addition to the above general summary, a fuller discussion of experimental methods and results will be found useful for the special student of this subject.

EXPERIMENTAL METHODS.

Cultures of Sporotrichum.—All our culture · experiments were made by the strict methods of the bacteriological laboratory, preliminary sterilization of the medium on which the fungus was grown and subsequent protection of the culture from bacterial invasion having been found necessary by experiments made in 1891. The media were either peptonized agar-agar, or a batter of raw corn meal made up with beef broth or other nutrient fluid prepared according to the customary procedure of the bacteriologist.

A few new experiments were made, additional to those of previous years, with variations in the culture medium, especially by acidulating it, and some preliminary trials were begun to determine the effect of variations in the temperature at which the cultures were kept.*

The apparatus used by us was in all cases either the common test-tube with a cotton plug, or a glass fruit jar of the " Mason " pattern (usually of a capacity of two quarts), the metal cap of which screws on to the top of the jar with a flat rubber ring intervening. The caps were altered by closely soldering a tin tube into an opening in the top of each (see Plate V., Fig. 1), as a safeguard against accidental infection by bacteria when the spores were sown upon the medium, and also for the purpose of convenient plugging with cotton as a subsequent protection.

In charging this jar with the culture medium, the metal cap was removed and the jar was partly filled with the corn-meal batter, mixed barely thick enough to settle smoothly, and was then placed upon its side, so that the mixture collecting at the lower part of the jar might present as large a surface as possible for the growth of the fungus. This culture jar worked very satisfactorily, any secondary infection of the culture rarely interfering with the growth and complete development of the Sporotrichum.

The cover of the jar was of course removed to get access to its contents, and if it was desired to preserve the culture for some time without deterioration the jars were left open until the contents were dried out. (See Plate VI.) It was found that such dried masses of

---

*Experiments of this description are much to be desired, so conducted as to determine the temperature *optimum* of *Sporotrichum globuliferum*, as well as the limits of cold and heat at which growth and spore formation come to a stand and at which the fungus itself, or its spores, may be destroyed. Similar experiments should be tried with the effects of various degrees of moisture.

corn meal with surfaces covered by Sporotrichum growths could be readily and successfully freshened and revived after some months by simply moistening the mass.

The various experimental cultures of the season were greatly interfered with by the pressure of more practical operations, and little was added to our previous knowledge of the subject. The growths on peptonized agar were invariably prompt and profuse, excepting where the medium was too highly acid, or where the temperature approximated 100° Fah. (See Nos. 29–35 and 48–50.) Growth from the spores at ordinary temperatures was commonly noticeable by the second day; sometimes not until the third. Heads of spores were visibly formed on the fifth or sixth day, and were ripe on the eighth day, at the earliest. It was found that the fungus grew more profusely and luxuriantly on an acidulated medium than on an alkaline or neutral one. (No. 29, etc.) A mixture of raw corn meal with water in which potatoes had been boiled was apparently better adapted for the culture of Sporotrichum than the batter of corn meal and beef broth, but this conclusion requires verification. The growth on these corn-meal mixtures was always at least as prompt and generous as on the agar gelatine.

Chinch-bugs in Boxes.—In our experiments with the transfer of muscardine to healthy chinch-bugs by enclosing them with specimens dead with disease in especially prepared boxes, we found that a layer of moist earth in the bottom of the box was an important aid to success, and that garden soil was better than sand. We had also abundant evidence that these experiments were most successful with weakened insects, and especially with those brought in from the field after the older generation present had passed its reproductive period and was consequently about to die. On the other hand, adults *in coitu* were occasionally found, one or both of which had died of muscardine.

Owing to unskilled methods of preparation and packing, and likewise to delays in transit, a large part of the material sent to the office was either dead when received or in a badly damaged condition. Although the worst of this material was always rejected, dead bugs accumulated so rapidly in our contagion boxes as to foul the contents, and to breed numerous blow-fly larvae and masses of Anguillulidae, and thus practically to interrupt the growth of the Sporotrichum. To avoid these disadvantages large reception boxes were prepared, each provided with a second bottom of coarse slats, a few inches above the first. The chinch-bugs received were placed on the lower bottom, and the vegetation used for food was laid upon the slats. When additions were to be made to the contagion boxes the stalks of corn and other food were taken out and beaten and shaken over the boxes, only the stronger and better-fed insects being thus transferred. While this procedure had the effect to eliminate the difficulties due to dead and rotting insects, it also brought the fungus development practically to a stand, and it was not until these more hardy chinch-bugs had been kept in confinement for some weeks that they began to suffer noticeably from muscardine.

The difficulties due to the appearance of mites in the infection boxes have already been referred to. Minute Anguillulidae, so abundant among dead chinch-bugs as to form gray patches here and there, did not seem to affect the fungus growth, neither were the blow-fly maggots especially injurious to these cultures so far as we could observe. On the other hand, as both these forms devoured dead chinch-bugs indiscriminately, they doubtless interfered with the development of the fungus in the boxes.

CLASSIFIED LIST OF EXPERIMENTS, 1894.

Nos. 1–91.*

CULTURES ON AGAR-AGAR.

No. 1, April 21.   Neutral culture from dead insect larva.
No. 6, June 28.   Neutral culture from No. 1.
No. 11, July 2.   Neutral culture from No. 1.
Nos. 29–35, Aug. 2.   Acid cultures from No. 11.
No. 43, July 3.   Neutral culture from No. 1.
No. 48, July 27.   Neutral culture from No. 43.   Temperature test.
No. 49, July 30.   Neutral culture from No. 43.   Temperature test.
No. 50, July 27.   Neutral culture from No. 43.   Temperature test.

CULTURES ON CORN-MEAL MIXTURES.

No. 2, May 7.   Neutral culture in fruit jar.   From No. 1.
No. 7, July 6.   Neutral culture on corn meal and agar gelatine. From No. 6.
No. 8, July 6.   Neutral culture on corn meal and potato water. From No. 6.
No. 9, July 6.   Neutral culture on corn meal and beef broth. From No. 6.
No. 10, July 9.   Acid culture from No. 1.
No. 12, July 13.   Neutral culture in fruit jar.   From No. 11.
No. 13, July 13.   Acid culture in fruit jar.   From No. 11.
No. 14, July 17.   Acid culture in fruit jar.   From No. 11.
No. 42, July 3.   Neutral culture in fruit jar.   From No. 1.
Nos. 44 and 45, July 20.   Acid culture in fruit jar.   From No. 43.

---

*These numbers correspond to those of the descriptions of experiments following.

CONTAGION AND INFECTION EXPERIMENTS WITH LIVING CHINCH-BUGS IN BOXES.

No. 3, May 11. From agar culture No. 1.
No. 4, May 16. From agar culture No. 1.
No. 5, May 17. From agar culture No. 1.
No. 54, May 25. First contagion box. Kansas chinch-bugs.
No. 56, June 7. Farmers' contagion box, G. C. Wells. From No. 54.
No. 68, June 22. First large laboratory contagion box. From No. 54.
No. 69, June 23. Second large laboratory contagion box. From No. 54.
No. 70, June 27. Third large laboratory contagion box. From No. 69.
No. 71, June 28. Fourth large laboratory contagion box. From No. 2.

LABORATORY INFECTION EXPERIMENTS WITH LIVING CATERPILLARS.

Cabbage worms (Pieris rapae).
Nos. 15 and 16, July 31. From acid corn-meal culture No. 14.
Nos. 18–20, July 28. From neutral agar culture No. 11.
No. 28, July 31. (Pupa.) From neutral agar culture No. 11.
No. 51, July 27. From neutral agar culture No. 43.
Elm-leaf caterpillar.
No. 17, July 28. Infected from neutral agar culture No. 11.

LABORATORY INFECTION EXPERIMENTS WITH DEAD INSECTS.

Chinch-bugs.
Nos. 36–40, Aug. 2. From neutral agar culture No. 11.
Nos. 46 and 47, July 20. From neutral agar culture No. 43.
Cabbage worms.
Nos. 21–24, July 30. From neutral agar culture No. 11.
Nos. 25–27, July 31. From neutral agar culture No. 11.
No. 41, Aug. 2. From neutral agar culture No. 11.
Sphinx larva.
No. 52, Aug. 2. From neutral agar culture No. 43.

FIELD EXPERIMENTS WITH MUSCARDINE FUNGUS.

No. 53, April 20. Hollenbeck farm, near Tonti.
Nos. 55 and 56, June 7. Wells farm, near Farina. From contagion box No. 54.

Nos. 57 and 58, June 10.   Wells farm, near Farina.   From contagion box No. 56.

No. 59, June 7.   Smith farm, near Farina.   From contagion box No. 54.

No. 60, June 15.   Smith farm, near Farina.   From contagion box No. 59.

No. 61, June 15.   University farm.   From contagion box No. 54.

No. 62, June 18.   University farm.   From contagion box No. 54.

No. 63, June 19.   Bartley farm, near Edgewood.   From contagion box No. 54.

Nos. 64 and 65, Aug. 6.   Bartley farm, near Edgewood.   From corn-meal culture No. 2 and contagion box No. 68.

No. 66, Sept. 4.   Bartley farm, near Edgewood.   From contagion box No. 68.

No. 73, Aug. 7.   Ferguson farm, near Odin.   From contagion box No. 68.

No. 74, Aug. 7.   Silver farm, near Odin.   From corn-meal culture No. 2.

No. 75, Aug. 7.   Robinson farm, near Odin.   From corn-meal culture No. 2.

No. 77, May 15 and June 10.   Heth farm, near Edgewood.   From contagion box No. 54.

No. 78, June 25.   Wilson farm, near Greenup.   From contagion box No. 54.

No. 79, July 1.   Jackson farm, near Greenville.   From contagion box No. 54.

No. 80, June 20.   Filson farm, near Xenia.   From contagion box No. 54.

SPONTANEOUS OUTBREAK OF MUSCARDINE.

No. 76, Oct. 6.   Hurd farm, near Odin.

See also No. 53, June 5; No. 55, June 7 and June 19; No. 57, concluding discussion; No. 60, June 19; Nos. 63–67, etc.

EFFECT OF MOISTURE ON CHINCH-BUGS.

No. 73, June 1.   Bugs confined in saturated air.

EXPERIMENTS WITH BARRIERS AND TRAPS.

No. 81, July 10.   University farm, furrow experiment.

No. 82, July 11.   University farm, furrow and post-hole experiment.

No. 83, July 12.   University farm, furrow and post-hole experiment.

No. 84, July 12.   University farm, coal-tar and post-hole experiment.

No. 85, July 13.   University farm, coal-tar and post-hole experiment.

No. 86, July 10.   University farm, furrow, coal-tar, and post-hole experiment.

No. 87, June 27.   Bartley farm, near Edgewood, field furrow experiment.

No. 88, June 28.   Smith farm, near Farina, furrow and log experiment.

No. 89, June 15.   Filson farm, near Xenia, furrow, post-hole, and kerosene emulsion experiment.

No. 90, June 23.   Mayo, near Falmouth, furrow experiment.

No. 91, June 25.   Wilson farm, near Greenup, furrow and log experiment.

OUTLINE OF EXPERIMENTS WITH CHINCH-BUG MUSCARDINE FUNGUS (*Sporotrichum globuliferum*, Speg.).

Nos. 1–80,*

APRIL 21 TO OCTOBER 10, 1894.

No. 1, April 21, agar culture, from dead insect larva.

No. 2, May 7, neutral corn-meal culture in fruit jars.

No. 64, August 6, field experiment, Bartley farm. (See also under 68.)

No. 65, August 6, field experiment, Bartley farm. (See also under 68.)

No. 67, September 7, field experiment, Bartley farm.

No. 71, June 28, large contagion box in laboratory.

No. 74, August 7, field experiment, Silver farm.

No. 75, August 7, field experiment, Robinson farm.

No. 3, May 11, laboratory infection experiment with chinch-bugs.

No. 4, May 16, laboratory infection experiment with chinch-bugs.

No. 5, May 17, laboratory infection experiment with chinch-bugs.

No. 6, June 28, agar culture.

No. 7, July 6, test-tube culture on corn meal and agar gelatine.

No. 8, July 6, test-tube culture on corn meal and potato water.

No. 9, July 6, test-tube culture on corn meal and beef broth.

No. 10, July 9, culture on acidulated corn-meal batter.

No. 11, July 2, agar culture in twelve test-tubes.

*The subordination and dependence of these experiments, one upon another, is indicated by the indentation of the items on this list. For example, all the material for the first fifty-two numbers was derived, directly or indirectly, from the dead insect larva referred to under No. 1; experiments 64, 65, etc., down to 75, were begun with Sporotrichum grown as stated under No. 2; the cabbage worms mentioned under Nos. 15 and 16 were infected from the acid culture No. 14, itself the third remove from the dead insect—and so on through the list. The numbers in this outline correspond, of course, to those used in the full description of experiments next following. It should be noted, however, that No. 72 and the barrier experiments, Nos. 81 to 91, are not represented in this list.

No. 12, July 13, fruit-jar cultures on neutral corn-meal batter.

No. 13, July 13, fruit-jar cultures on acidulated corn-meal batter.

No. 14, July 17, fruit-jar cultures on acidulated corn-meal batter.

No. 15, July 31, infection experiment with live cabbage worms.

No. 16, July 31, infection experiment with live cabbage worms.

No. 17, July 28, infection experiment with elm-leaf caterpillar.

Nos. 18–20, July 28, infection experiments with live cabbage worms.

Nos. 21–24, July 30, infection experiments with dead cabbage worms.

Nos. 25–27, July 31, infection experiments with dead cabbage worms.

No. 28, July 31, infection experiment with pupae of cabbage worms.

Nos. 29–35, Aug. 2, test-tube cultures with acidulated agar.

Nos. 36–40, Aug. 2, infection experiments with dead chinch-bugs.

No. 41, Aug. 2, infection experiment with dead cabbage worms.

No. 42, July 3, fruit-jar culture on corn-meal batter.

No. 43, July 3, test-tube agar culture.

Nos. 44 and 45, July 20, fruit-jar cultures on acidulated corn-meal batter.

Nos. 46 and 47, July 20, infection experiments with dead chinch-bugs.

No. 48, July 27, agar culture, temperature test.

No. 49, July 30, agar culture, temperature test.

No. 50, July 27, agar culture, temperature test.

No. 51, July 27, infection experiment with cabbage worms.

No. 52. Aug. 2, infection experiment with sphinx larva.

No. 53, April 20, field experiment, Hollenbeck farm.

No. 54, May 25, laboratory contagion box, Kansas chinch-bugs.

No. 55, June 7, field experiment, Wells farm.

No. 56, June 7, contagion box and field experiment, Wells farm.

No. 57, June 10, field experiment, Wells farm.

No. 58, June 10, field experiment, Wells farm.

No. 59, June 7, contagion box, field experiment, Smith farm.

No. 60, June 15, field experiment, Smith farm.

No. 61, June 15, field experiment, University farm.

No. 62, June 18, field experiment, University farm.

No. 63, June 19, field experiment, Bartley farm.

No. 68, June 22, large contagion box in laboratory.

No. 64, Aug. 6, field experiment, Bartley farm. (See also under No. 2.)

No. 65, Aug. 6, field experiment, Bartley farm. (See also under No. 2.)

No. 66, Sept. 4, field experiment, Bartley farm.

No. 73, Aug. 7, field experiment, Ferguson farm.

No. 69, June 23, large contagion box in laboratory.

No. 70, June 27, large contagion box in laboratory.

No. 77, May 15 and June 10, field experiment, Heth farm.

No. 78, June 25, field experiment, Wilson farm.
No. 79, July 1, field experiment, Jackson farm.
No. 80, June 20, field experiment, Filson farm.
No. 76, Oct. 6, spontaneous outbreak, Hurd farm.

DESCRIPTION OF EXPERIMENTS.

1. Experiments with the Fungi of Contagious Disease.

The season's operations with contagious-disease experiments were developed principally from two points of departure: the first a single insect larva, indeterminable as to species, found April 17 dead and covered with the fungus of white muscardine, in a plowed corn field near Urbana; and the second a small lot of chinch-bugs dead with the same fungus infection, received from Chancellor Snow of the University of Kansas about May 15. From the first-mentioned material numerous cultures were made, and infection experiments were conducted on chinch-bugs, cabbage worms, and other insect larvae, the series of operations dependent upon this original specimen extending from April 21 to September 20. From the second lot various laboratory and field experiments were started—twenty-one in number—extending from May 25 to October 10, all having the character of direct or indirect exposures of living chinch-bugs to contagion or infection by means of these dead specimens obtained from Dr. Snow.

Although several of the experiments here described were incomplete or otherwise unsatisfactory, I have thought it best to report the whole mass of them, as an assurance (if for no other reason) that nothing has been withheld because of its unsatisfactory character.

No. 1. April 21. A test-tube culture on agar, from a dead insect larva of undetermined species, found in a field at Urbana, Illinois. April 24, had barely begun to grow. April 27 spores had formed. April 30, spores ripe, the culture unmistakably *Sporotrichum globuliferum.* This larva and the culture derived from it are the starting point for all experiments of this list to No. 52 inclusive, as well as for Nos. 64, 65, 67, 71, 74, and 75 additional.

No. 2. May 7, six Mason fruit jars, with caps altered to facilitate sterile culture (see Plate V, Fig. 1), partly filled with a batter of corn meal and beef broth and sterilized by dry heat for one hour on each of two successive days, were inoculated by spores of the muscardine fungus taken from culture No. 1. This second culture was successful, and furnished material for a large amount of subsequent experimental work.

No. 3. May 11, a lot of chinch-bugs received from F. O. Pierce, of Xenia, Clay .county, was treated with spores from culture No. 1 and placed in a small wooden box which was kept on wet earth under a hedge, and covered again with a box of larger size. May 16 these bugs began to die. By the 22d all were dead but one, and an external growth of the Sporotrichum had begun to appear upon three. All were then returned to the sender for distribution in his fields.

No. 4. A precisely similar experiment with chinch-bugs from Trenton, Illinois, was begun May 16, the experimental box being similarly placed. On the 19th some of the bugs were dead; on the 22d an external fungous growth appeared upon two of them, as yet, however, without spores; on the 25th many more were dead, several of them covered with a white mycelium; and by the 29th nearly all had perished, development of the characteristic spores on several of them now giving unmistakable evidence of the presence of *S. globuliferum*—the special fungus of the muscardine disease. Specimens returned to the sender.

No. 5. On the 17th of May a second lot of bugs from Xenia was similarly treated, with a similar result. Beginning to die May 19, most of the bugs were dead on the 22d, and on the 29th all had perished. A slight external growth appearing at this date resembled in every way the immature mycelium of the muscardine fungus. Without waiting for further evidence of infection this lot of bugs was returned to the sender.

No. 6. Next, on the 28th of June, nearly two months from the original agar culture (No. 1), a second agar tube was infected from that growth. July 6, this culture was ripe, and the spores were used for the experiment next succeeding.

No. 7. July 6 a test-tube mixture of corn meal and agar gelatine was infected from culture No. 6—the third remove from the dead insect. July 9 this had made a good start, but was not followed further.

No. 8. July 6. This was an experiment identical with No. 7, and begun at the same time, varying only in the culture medium used, which was a batter of corn meal mixed with water in which potatoes had been boiled. Three days later, July 9, the spores were starting abundantly, but the matter was not followed further.

No. 9. July 6. This was a companion experiment to the two preceding, except that our ordinary mixture of corn meal and beef broth was used in place of the foregoing media. July 9, growth had begun, but less vigorously than on either of the others.

No. 10. July 9. A corn-meal batter, like that of No. 9, but acidulated with acetic acid, was infected with ripe spores from No. 1. July 11 it had made a good start, but was not followed further.

Nos. 7 to 10, derived from No. 6, were intended originally to test the comparative value of various corn-meal mixtures. The exigencies of the season's work prevented their completion, and they serve merely to give some hints suggestive of further trials.

No. 11. July 2. This was a repetition of No. 6—a set of agar test-tube cultures, twelve in all, the spores for which were derived from No. 1. July 5 all were freely growing, and July 9 the growth was spreading rapidly. These tubes were not again reported on until August 13, at which time an abundant development of spores of the muscardine fungus was noted on all, the gelatine having, however, in the meantime quite dried up. A long series of cultures on corn meal and of infection experiments on chinch-bugs and cabbage worms, and other larvae, were made with spores from this set of tubes.

No. 12. July 13, 1:30 p. m., several fruit jars of the corn-meal mixture with beef broth, neutralized with sodium carbonate, were infected with spores from No. 11. Two days later, July 15, these spores had begun to grow, and on the 17th the growth covered a part of the surface.

No. 13. July 13, at 1:30 P. M., spores from No. 11 were sown on acidulated mixture of corn-meal and beef broth which had been placed in altered Mason fruit jars and sterilized by heat as already described. On the 15th growth had begun, and on the 17th the surface of the meal was covered.

No. 14. July 17, at 2 o'clock P. M., fourteen Mason fruit jars with the acid mixture of corn meal and beef broth, sterilized by heat on two successive days, were infected with spores from No. 11. On the 19th the growth had started, and on the 21st it had become very abundant, distinctly more so indeed on both these acid mixtures (13 and 14) than on the neutral mixture No. 12.

No. 15. July 31. An infection experiment on living cabbage worms with spores from the acid corn-meal culture, No. 14 of this list. Two cabbage worms were infected on the back near the head, and were then shut up in a large covered glass dish which had a layer of moist sand on the bottom, on which a fresh cabbage leaf was placed for food. August 1, no growth has appeared on either cabbage worm. August 2, still no growth, but one of the larvae has pupated. August 3, still no growth. August 7, one larva dead, softened, and blackened, but without appearance of fungus contamination. The pupa also dead and softening. August 13, no Sporotrichum apparent, only softened bodies of larva and pupa remaining.

No. 16. July 31. An experiment parallel with the preceding, three cabbage worms being infected on the back. Results identical with those of No. 15, one worm pupating August 2. Both larvae and pupa died, blackening and becoming deliquescent, without appearance of muscardine. The pupating larva proved to have been parasitized, as is noted August 7.

These two experiments with infection material derived from acid cultures throw some doubt on the effectiveness of that kind of material, but as the larvae evidently died from the common bacterial disease of the cabbage worm, it is possible that the previous presence of this disease prevented the development of the fungus, an hypothesis that is made more probable by the fact that spores of Sporotrichum will not germinate on decaying media.

No. 17. July 28. An infection experiment upon the larva of a butterfly (*Grapta interrogationis*) taken from the elm. This caterpillar was infected with spores from agar culture No. 11, itself derived, as will be remembered, from agar culture No. 1, derived in turn from the dead larva with which this series began. The material here used was consequently at two removes from the dead insect. The spores were placed along the center of the back of the caterpillar at 10:30 A. M., the infected insect being then shut up in a large covered glass dish, with a layer of moist sand on the bottom, upon which leaves of elm were scattered as food. At 9 A. M., July 30, this larva had pupated, of course casting off the skin upon which the spores had been placed. It was examined daily without note of change until August 7. The pupa at this date was still alive and apparently healthy, but exhibited a slight growth of Sporotrichum from a point about the size of a pin head on the ventral surface, at the edge of the pupal wing-pads. This ventral point was in immediate contact with the cast skin still bearing the fungus spores. August 13 this pupa was dead, with a slight external growth of *Sporotrichum globuliferum*—at that time in fruit.

No. 18. July 28, 10:00 A. M. An infection experiment similar to the preceding, except that a cabbage worm (*Pieris rapae*) was used instead of the Grapta larva, and that a cabbage leaf was placed in the dish instead of leaves of the elm. The caterpillar was touched with spores from No. 11 on the back, immediately behind the head. July 30 the fungus was growing freely from the point of infection and spreading to the sides of the body. July 31 it had enveloped the body near the head, and had also extended on the ventral surface the entire length of the caterpillar. The color of the larva had in the meantime changed to that of crushed raspberries wherever the fungus was growing, and August 1 the entire cabbage worm was of the same crushed-raspberry color. August 2 the Sporotrichum had spread all over the surface, and August 7 it was well developed everywhere and covered with ripe spores.

No. 19. July 28. An infection experiment precisely like the last, except that the spores were placed along the right side of the caterpillar only. Two days later the fungus growth was abundant all along this side wherever the spores had lodged, but it had not yet begun to spread. A very fine white web had fastened the larva to the dish upon which it was resting when it died. July 31 nearly the entire body was covered by the mycelium of Sporotrichum, leaving only the two ends free, and the larva had begun to change to the crushed-raspberry color. August 2 the whole body was covered, and later the spores developed everywhere, as before.

No. 20. July 28. Like the preceding, except that the spores were applied only at the posterior end of the back. July 30, growth abundant at the point of infection, covering about a fourth of the back, but not extending downwards to the sides. A very fine white web fastens the larva to the cover July 31, growth of mycelium very abundant at point of infection, and extending downwards, underneath the body, to the hinder end, the crushed-raspberry color appearing wherever the fungus has taken hold. August 1, growth slowly extending downwards. August 7, spores well developed on the dead larva.

Experiments 15, 16, 18, 19, and 20 of the foregoing series show the efficiency of this fungus as a means of infecting living cabbage worms, and bring to light also the interesting fact that its growth may have the effect to color the larva red.

We have next a series of experiments intended to test the possibility of the growth of Sporotrichum on insects dead when infected. For this purpose, cabbage worms killed with chloroform were used.

No. 21. July 30, 1:30 P. M. Chloroformed cabbage worm placed on piece of cabbage leaf on moist sand, spores of muscardine (Sporotrichum) from No. 11, sown along the back. July 31, 8:30 A. M., no growth visible. August 1, 9 A. M., slight growth along the back. August 2, 8:30 A. M., growth very abundant on the back. August 7, 3 P. M., Sporotrichum well developed on this larva. Specimen preserved.

No. 22. July 30. This was an experiment like No. 21, except that the cabbage worm was infected on the ventral surface. July 31, growth not started. August 1, beginning to grow. August 2, growth very abundant and spreading over the entire body. August 7, Sporotrichum well developed. Specimen preserved.

No. 23. July 30. Same as the foregoing, but infected along the side. Examined August 1, 2, 3, and 7, with results exactly as above.

No. 24. July 30. Exactly as above, except that several caterpillars were infected on the head. July 31, growth not started. August 1, no growth as yet, except that a fine cobweb fungus has started on a single worm. August 2, the above worm completely enveloped in the cobweb growth. August 3, germination of spores just beginning to show. At this point the above experiment was accidentally interrupted.

Nos. 25, 26, and 27. July 31. Dead cabbage worms as above. Experiments identical with the foregoing, and made at the same date, except that infected spots were touched with sterilized distilled water to cause the spores to adhere to the worm. The results were all as above mentioned, with a slight variation in No. 27. Here, August 1, no growth of the spores had appeared, but one larva had become fastened to the cover by a fine white web. On the 2d the growth was spreading feebly, and on the 3d the worms were turning crushed-raspberry color, the one on the cover having the Sporotrichum growth well started. August 7 this fungus was fully developed on all, and the specimens were put aside for preservation.

No. 28. July 31, 10 A. M. Three pupae of cabbage worms placed under same conditions as No. 21, and infected with Sporotrichum from No. 11. August 1, no growth. August 2, two of the pupae turning reddish, but no growth. August 3, still no growth, but one of the pupae of a deep red tint. August 7, two of the pupae dead, one of them showing a slight mycelial growth of Sporotrichum. The third has yielded a healthy butterfly.

Nos. 29 to 35 are agar test-tube cultures, in which the gelatine was acidulated variously, it being the purpose of these experiments to determine the degree of acidity of the medium most favorable to the germination of the spores and the development of the fungus.

No. 29.   August 2, 3¾ cc. of agar gelatine and ⅛ cc. of a one-eighth per cent. solution of acetic acid in test tube, sown with spores of *Sporotrichum globuliferum* from No. 11.   August 13, no Sporotrichum growing, spores apparently not having started.

No. 30.   August 2.   Like 29, except that the acetic acid was in a one-fourth per cent. solution.   August 13, there has been apparently a slight growth of Sporotrichum in this tube, but it is now all dead.

No. 31.   August 2.   Like 29, except that the acetic acid used was a one-half per cent. solution.   August 13, a small straggling growth of Sporotrichum has appeared and has formed spores.

No. 32.   August 2.   As above, except that the acetic acid was a one per cent. solution.   August 13, fair growth of Sporotrichum, covering most of the agar surface. Spores abundant, ripe, but not of yellowish tint.

No. 33.   August 2.   As above, acetic acid now a two per cent. solution.   Result identical with No. 32.

No. 34.   August 2.   Like No. 29, except that the solution of acetic acid is now five per cent.   August 13, no growth whatever.

No. 35.   Like No. 29, except that the acetic acid used was a ten per cent. solution. August 13, no growth.

From the foregoing it appears that an agar mixture containing from three to six hundreths of one per cent. of acetic acid is that which proved most favorable to the growth of Sporotrichum.

Next, I report six experiments with dead chinch-bugs, intended to test the possibility of the infection of this insect, after death by other causes.

No. 36.   August 2.   Fifteen chinch-bugs killed by crushing were placed in a covered glass dish, and treated with spores from culture No. 11.   August 3, no appearance of growth.   August 7, the dish a mass of mould, which fills the entire interior; no appearance of a growth of Sporotrichum on the bugs.   August 13, no development of Sporotrichum in this experiment.

No. 37.   August 2.   A duplicate of No. 36, with precisely the same result.

No. 38.   August 2.   Chinch-bugs killed with chloroform, placed on a cabbage leaf on moist sand in a covered glass dish and infected with Sporotrichum from No. 11. August 3, no growth.   August 7, no appearance of infection; specimens beginning to mould.   August 13, no Sporotrichum to be found.

No. 39.   August 2.   Equivalent experiment, except that the bugs were placed immediately on sand.   No growth, August 3.   August 7, 3 P. M., slight appearance of Sporotrichum on legs, antennae, etc., of single insects, here and there.

No. 40.   August 2.   The same as 38.   August 3, no growth.   August 7, loosely covered with bluish mould; no Sporotrichum formed.

No. 41.   August 2.   Infection experiment with dead cabbage worms.   Equivalent of No. 25.   August 3, no growth.   August 4, development of Sporotrichum commencing on all these cabbage worms.   August 7, Sporotrichum well grown, with some appearance of *post mortem* mould.

No. 42.   July 3.   A fruit-jar culture on corn-meal batter made with beef broth, the spores for which were taken directly from agar culture No. 1.   A successful growth, not requiring detailed report.

No. 43.   July 3.   An agar culture in five test-tubes, made, like 42, from No. 1. Developed readily and matured in due season, being subsequently used for a considerable number of cultures and infections.

No. 44.   July 20.   Fifteen jars of the acidulated corn-meal mixture (degree of acidity not given) infected with Sporotrichum from No. 43 at 1:30 P. M.   These cultures were all well started July 23, and matured in due season, eight of the fifteen being, however, somewhat contaminated by Aspergillus. The contents of these jars were subsequently used in part for chinch-bug infection experiments on the large scale.

No. 45. July 20. Culture like 44, differing only by an abbreviation of the sterilization process. Previous to this the jars had first been sterilized by heat while empty, then stocked with the corn-meal mixture and sterilized again. In the present culture, jars and batter were sterilized together, once for all. Result the same as 44.

No. 46. July 20. Chinch-bugs killed by chloroform were placed dry in a test-tube, treated with Sporotrichum from 43, and left with the test-tube plugged with cotton. July 28, no fungus growth. ·

No. 47. July 20. Like 46, except that the chinch-bugs were moistened with distilled water before being placed in the tube. July 28 and August 13, no growth at either date.

July 27, a beginning was made with experiments intended to determine the temperature at which Sporotrichum would grow most freely.

No. 48. July 27, 11 A. M. An agar test-tube sown with spores from No. 43 and kept in laboratory at temperature of 80° Fah. July 28, 8.30 A. M., 74° Fah.; 11:30, 79° Fah. July 30, 8:15 A. M., 74° Fah., growth starting nicely. The average temperature by day in this culture was about 77°.

No. 49. July 30. Another tube, as above, placed in south window, temperature ranging from 77° to 95° Fah. July 30, growth started well.

No. 50. July 27. Agar tube, as above, placed in incubator at 97° Fah. at 11 A. M., and kept there at a constant temperature of 100° Fah. July 30, growth not starting. August 13, these spores did not germinate.

No 51. July 27. An experiment with cabbage worms infected from No. 43. Several caterpillars (*Pieris rapae*) placed under a sterilized glass bell-jar with a large piece of fresh cabbage leaf lying on table without moist sand. Spores were shaken from agar culture and spread upon the worms with a sterilized platinum wire. July 30, several worms have died, but without appearance of fungous growth. One example, however, shows growth near the hinder end, and has turned black about one-fourth of its length. July 31, the cabbage worm just mentioned has turned a crushed-raspberry color, and the Sporotrichum has begun to spread over the surface. Another larva has turned a similar color, and a Sporotrichum growth is appearing at one point. August 1, several of the worms have become crushed-strawberry color, but without visible growth. One has become brownish-green, and on this Sporotrichum filaments have appeared externally. August 7, everything dead but one imago recently emerged. Ten died in pupa stage; two larvae parasitized. August 13, very little external development of fungus has appeared in this lot of worms.

No. 52. August 2. Large sphinx larva (Protoparce) chloroformed, and placed in a large culture dish upon a cabbage leaf resting on moist sand. Spores from culture 43 sown upon the right side. August 3, no growth. August 7, mixture of fungi all over surface, but infected area with a conspicuous white patch. August 13, no *Sporotrichum globuliferum*, only common mould.

No. 53. This is a farmer's contagion experiment made by Mr. C. S. Hollenbeck, near Tonti, Illinois, and not observed by us in the beginning. Infection material originally obtained from my office September 7, 1893, and used to start contagion box, the contents of which were afterwards distributed in his fields with a result which he regarded as successful last fall. The box, with a considerable number of chinch-bugs remaining, was kept over winter in a warm and rather moist cellar, and about April 20 its contents were scattered in wheat in a young orchard, care being taken to place the material where bugs were thickly congregated. The weather at the time was damp, cloudy, and warm. Mr. Hollenbeck reports that about a week afterwards he saw many mouldy bugs in the field, other than those put out, and was convinced that these bugs were killed by fungus disease.

May 24 this wheat was examined by Mr. Marten, who reported the chinch-bugs very abundant in the wheat, copulating and depositing eggs   Search for the greater part of a day on this farm and one adjacent, yielded only two fungus-covered chinch-bugs.   On another visit, made June 5, the fungus disease was found generally distributed in this and adjoining fields, as it was at that time throughout this part of the state at large; but June 20, after an interval of drought, no trace of it could be found by Mr. Marten in these same fields.   The young bugs at this time were plentiful, a few having reached the pupa stage.

No. 54.   A contagion experiment, the first of a long series (see Outline, page 40) derived from a small lot of chinch-bugs, dead with *Sporotrichum globuliferum*, received from Dr. Snow about May 15.   May 25 a lot of chinch bugs from the vicinity of Tonti and Odin, in Marion county, collected by Mr. Marten, together with others received from fourteen farmers of that vicinity, were placed in a wooden box which had been thoroughly wet inside and out, and the bottom of which was covered with a layer of green wheat for food.   With these bugs the material obtained from Dr. Snow was placed.   The box so prepared was tightly closed and kept on damp sand upon the ground in the insectary.   It was opened for examination about every other day and supplied with fresh food.   May 26 more bugs were added, but no dead insects were observed.   May 28 another lot from southern Illinois was introduced, and still another on the 29th.   No dead were seen at this time, and there were no traces of muscardine infection.   On the 30th a few bugs were dead in the box, two of them well covered with another chinch-bug fungus, *Entomophthora aphidis*, but no fresh Sporotrichum was seen.

More live insects were introduced May 31 and June 1, at which latter date the dead bugs were a little more numerous.   June 4, the box had warped and split, and more than half the bugs escaped; otherwise the experiment was in good condition.   The white muscardine had now taken effect, and a sufficient number of fungus-covered chinch-bugs were taken out to supply the fourteen farmers near Tonti and Odin from whom the material was received May 25, together with four other farmers at Farina, in Fayette county.   Among these last were G. C. Wells and James Smith, whose farms were frequently visited by us later.   The lot sent to Tonti and Odin was delivered June 5, and that to Mr. Wells and Mr. Smith June 7, a part of this latter material being used by Mr. Marten for the field experiment on the Wells farm, reported at length under No. 55.

June 9 all the material was taken from the above box and placed in a second similar one.   June 11 this second box was overhauled and a sufficient number of fungus-covered bugs was removed to make up twenty packages, two lots of living bugs being at the same time added.   June 12 and 13 more live bugs were added, and a few dead ones were taken out.   June 14 the condition of this lot was regarded as unsatisfactory, and everything was again transferred to a clean box.   Of those which were dead a few were covered with muscardine fungus, but at least a hundred times as many gave no external trace of muscardine.   Five additional lots from various places were now introduced, and the box was a second time infected, by means of fifty chinch-bugs covered with ripened Sporotrichum collected by Mr. Marten in wheat fields at Tonti and Farina June 6 and 7 (see No. 55).   June 15 thirteen lots more were placed in this box, and a quantity of fungus-covered dead were removed and distributed in spring wheat on the University experimental farm (see No. 61).   June 16 a quantity was taken out for shipment, and seventeen packages from farmers were introduced.   June 17 six lots more were placed in this box, and June 18 four more   On this last date both old and young were dying with white muscardine, and a few with Entomophthora.   About three thousand live bugs from this box were now distributed in spring wheat on the University farm (see No. 62), and a lot of dead bugs removed and prepared for shipment, together with a quantity of both dead and living for use in fields in southern Illinois (see No. 63).   June 20, box overhauled, dead bugs removed and distributed to farmers. June 22, enough material removed to supply fifty-one farmers, after which the entire contents were transferred to two large boxes in Natural History Hall (see Nos. 68 and 69).

No. 55. June 7. A field infection experiment, started by Mr. Marten in a 4½-acre wheat field on the Wells farm, represented at A, Plate I. Chinch-bugs abundant, literally covering the wheat in many places, especially in the northeast corner adjoining corn (D). A dozen bugs dead with the white fungus, collected from the ground in wheat at this time, were afterwards placed in No. 54. Diseased insects locally present in all fields examined. About one hundred fungus-covered chinch-bugs from No. 54 distributed along the second and third drill rows, at the bases of plants where live bugs were most numerous, for a distance of several rods on the north and east sides, in the. northeast corner (indicated by the heavy dotted lines on Plate I).

June 19, field examined by Mr. Marten. Fungus-covered bugs about as abundant throughout the wheat as on the former visit. No indication that the disease had spread, or was any more prevalent in the vicinity of the spot where the dead bugs were distributed than at other places. Many fields examined in this neighborhood not entered on former visit. Diseased bugs found in all in moderate numbers. A second lot of infection material, consisting of several hundred bugs dead with this fungus, from No. 54, scattered on the ground in several drill rows in the immediate vicinity of those previously distributed. Wheat badly damaged and considerably lodged in several places. Bugs everywhere abundant, advancing into corn (B and D). June 20 Mr. Marten revisited the field (A) to examine two spots (*a* and *b*), each about one rod in diameter, where manure had been piled previous to being scattered, and where the wheat was badly lodged. The ground was here quite damp, and chinch-bugs had collected in considerable numbers. The muscardine fungus was also much more abundant here than elsewhere in the field, and several hundred whitened bodies could have been collected. It was, however, thought advisable to allow them to remain, in order to determine, if possible, whether or not the disease would spread from these centers of spontaneous development. Wheat cut June 25, but so far as could be seen, according to Mr. Wells, the fungus had not spread.

July 11 a few traces of the original material were found by Mr. Marten in the northeast corner.• A few whitened bodies were under the fallen wheat at *a* and *b*, but the fungus was apparently a week or ten days old, and there was no indication that the disease had spread from these places. Only an occasional live insect seen in the stubble, the great majority having migrated into adjoining corn, B and D, where considerable mischief was done. Several rows in the latter field had already been killed, and the corn was blackened with bugs for several rods. In the former field, however, these bugs were not quite so abundant, their progress being checked by a narrow lane 1 rod wide and an orchard (E) 4 rods wide, thickly grown up with weeds and grass, which separated A from B. The chinch-bug hordes were constantly emerging from the grass, however, and entering the corn. Many were hiding under clods and rubbish in both fields, but not one was found dead with any kind of fungous growth.

Mr. Wells reported August 8, that, in his opinion, we had wasted our labor, as he could not find any indication that the disease had spread to any part of his farm.

No. 56. A farmer's contagion experiment conducted by Mr. Wells, and examined several times during the season by Mr. Marten and Mr. Johnson. The box, 12x18x6 in., was prepared according to our directions (see p. 28), with a bottom layer of dirt half an inch deep, moistened and covered with a layer of corn leaves and green wheat. About two dozen diseased chinch-bugs from No. 54 were placed in the box June 7, together with a little more than a quart of live insects collected from wheat and corn in the neighborhood of No. 55. The box thus stocked was tightly closed, covered with a wet grain sack, and kept in the cellar on the damp floor.

June 10, nearly a quart of insects, both dead and alive, were removed from the box, several hundred of the dead being well covered with the fungus. All the insects, both dead and alive, were used for starting infection experiments in corn and oats (see Nos. 57 and 58). Several dozen whitened bodies were left in the box, into which about one pint of live insects collected from corn (B) were put, together with a fresh supply of corn leaves for food. The box remained in good condition and was supplied with fresh food and live insects about every third day. June 20 the second lot of material was removed,—about three pints in all,—consisting, as before, of dead and living insects. Box examined by Mr. Marten and found in excellent condition; several dozen bugs dead and covered with a fresh fungus growth left in the box; fresh food and about one-half pint of live insects introduced.

Box supplied with live insects and fresh food as needed. June 30, condition of box about the same as above. About one quart of chinch-bugs removed and placed in corn (B). Fresh food and one pint of live insects introduced. The box was left undisturbed, except when fresh food and live insects were put in, until the latter part of July. On one occasion, about the middle of July, Mr. Wells found a large insect, which he took to be a cockroach, under the box, "thickly covered with the same white fungus." It was removed and placed behind a leaf sheath on corn (B) among "a tea-spoonful of chinch-bugs," and left several days, without any indication that the disease spread to the insects coming in contact with it. It was afterward placed behind another leaf, similarly covered with chinch-bugs, where it remained; but at no time were there any traces of the fungus on the insects about it.

August 1, the box in good condition, and several thousand fungus-covered bugs removed, together with nearly three-fourths of a quart of live insects, all of which were placed in corn (B), as recorded under the following number. Fresh food and about one gill of live insects added; the box set away and not examined again until September 19, at which time it was overhauled by Mr. Johnson. Insects all dead; the dirt in the bottom somewhat dry, and the food considerably moulded; but the whitened bodies of dead chinch-bugs were very abundant. Several thousand well-covered specimens could have been taken out, but the supply was reserved for further use.

Mr. Wells informed Mr. Marten August 8 that his contagion box did not work so well after he began putting in immature bugs; but that, altogether, he had distributed in corn about four quarts of chinch-bugs that had passed through his box.

However successful this experiment may seem to have been, it must be borne in mind that the chinch-bugs introduced from time to time had been liable to infection in the fields where they were collected (see Nos. 55 and 57), as well as in the contagion box itself.

No. 57. A farmer's field infection experiment made by Mr. Wells on his farm near Farina, in corn (B). The first lot of material, about two-thirds of a quart of chinch-bugs, dead and alive, from his contagion box (No. 56), was distributed June 10 behind the leaf sheaths and on the ground of the first, third, and fifth rows along the south and east sides, where the bugs covered the first five rows of corn. This field was examined by Mr. Marten June 19, but no dead bugs were seen, and only a few traces of the original material were found. The chinch-bug attack was spreading rapidly, and corn was badly damaged throughout the south and east parts of the field.

The second distribution was made June 20, the material consisting of about three pints of chinch-bugs, dead and alive, from No. 56. The first three rows on the south, east, and west sides were treated as above. Examined by Mr. Wells June 25. No indications that the disease was spreading.

A third distribution was made June 30. About one quart of chinch-bugs, dead and alive, from same source as the others were scattered, as above, in the second, third, and fifth rows on the west and north sides. Mr. Wells examined field July 5, but found no bugs dead with the fungus, save a few scattered fragments of the original material. Mr. Marten examined this(B), as well as neighboring fields (D and F), July 11. No traces of the disease were found at this time on this farm, except a few weathered specimens in wheat (A), as noted in experiment 55. All the corn on this farm was dwarfed and ragged. The bugs were everywhere abundant, covering the stalks in many places throughout the fields.

The fourth, and last, distribution was made August 1. Three-fourths of a quart of chinch-bugs, dead and alive, from No. 56 were distributed, as above, along the west side in the third, fifth, and seventh rows, and in three alternate rows through the center of the field. Mr. Wells examined the field August 5, and reported that he could find no signs of the fungus other than a few traces of the original material.

From the above it will be seen that nearly four quarts of chinch-bugs, thoroughly exposed to the disease, were scattered in corn (B), two distributions each being made on the south, east, and west sides, one on the north, and one through the center of the field, as represented by the heavy dotted lines on the plate. Nevertheless this fungus practically disappeared from this farm during July and August, and did not appear again until about the middle of September. The chinch-bugs, however, continually increased in numbers, and caused far greater loss to corn than the drouth in this vicinity. September 4 Mr. Johnson found a few traces of the original material along the south and east sides of the field (B). No fungus found in corn marked D, although live bugs were very abundant throughout the field. Nothing indicating the presence of the white fungus was seen in corn marked F, or in any other field on this farm, except the one (B) just mentioned. A few bugs were found in all the meadows and pastures adjoining corn, but all traces of the local fungus-outbreak which had appeared in June had now utterly disappeared.

Nothing further of importance was noted from this neighborhood until September 18, at which time Professor Forbes received the following interesting letter from Mr. Wells: "I have been cutting corn and find diseased bugs scattered over the ground, especially under stalks lying on the ground. I have seen 30 or 40 under a single ear. I have .worked with this disease with your assistants, and know what I am talking about. There are still hordes of live bugs in the field."

The field referred to proved to be that marked B, and was carefully examined by Mr. Johnson September 19. The corn had been cut and shocked, the work having been finished the previous day. The ground was rather damp, but not muddy except in the southeastern part, which is somewhat lower than the rest of the field. There had been a heavy rain September 4, and according to Mr. Wells a slight shower had fallen about September 11, followed by a heavy rain again on the 16th. The east-central, southern, and southeastern parts of the field were rather weedy, and the corn had fallen here much more than elsewhere. The weeds, broken corn stalks, and leaves almost completely covered the surface of the ground, which was quite damp and rather sticky in such places. Chinch-bugs had collected here in considerable numbers, and whitened bodies covered with Sporotrichum were very conspicuous behind leaf-sheaths, on the stubble, on the surface of the ground in the open field, under clods, bits of fallen leaves, sticks, and rubbish of all kinds; on the ground under weeds and grasses; in corn shocks, on the stalks, behind the leaves; and on the ground under the shocks. They were, in fact, generally distributed throughout the field, being most abundant in those portions thickly covered with weeds, corn, etc., as indicated above. At a point, *c*, under a dense cluster of weeds where the ground was quite damp, Mr. Johnson collected 157 fungus-covered bugs from a surface area of two square feet; 68 were counted at *d* within a radius of ten inches, and 39 were found under a single shock at *e*. Similar examples were reported from all parts of the field. The great majority of the insects attacked by this fungus were adults, although several young of the first and second moults were seen.

Live chinch-bugs were abundant throughout the stubble, and had accumulated in great numbers in the shocks; but there was a general movement in all directions into adjoining meadows and pastures. The larger portion were in the pupa stage, although all ages were seen, even those just emerged from the egg.

No traces of the disease were found at this time in the wheat stubble (A), and only an occasional bug dead with the fungus was seen in corn D. Seven whitened bodies were collected in the vicinity of *f*. The entire northwestern part of this field had been invaded by chinch-bugs from wheat A, and was seriously injured; other portions suffered less damage.

In an eight-acre field of corn (F) to the north of D and east of B, the chinch-bug injury was very much more complete than in the latter, and about the same as in the former. Dead bugs covered with Sporotrichum were very numerous, being almost, if not quite, as abundant as in B. The corn had not been cut, and the diseased bugs were found behind leaf sheaths, under fallen stalks on the ground, around the hills and between the rows, and under clods, leaves, weeds, grasses, and other rubbish. Seventy-eight whitened bodies were counted on the ground under a single fallen stalk near the center of the field. Live bugs, mostly adults and pupae, were every-where abundant, and completely blackened the corn in many places, especially through the central and east-central parts. According to Mr. Wells' estimate, one-third of the entire field had been completely ruined at this time.

It must be borne in mind that not a single diseased insect was artificially intro-duced into this field at any time during the season, and that no traces of the fungus were found by Mr. Johnson at the time of his former visit, September 4. It seems quite possible, therefore, that the spores of this fungus were quite as abundant in F as in B, and that the parasite developed spontaneously in both fields when conditions fostered the growth of the germ.

All other corn fields within a radius of three-quarters of a mile were examined, but nothing was seen indicating the presence of the fungus. All the corn seen was badly damaged, and chinch-bugs were still very numerous.

Mr. Johnson examined the fields in this neighborhood again September 28. The afternoon was very warm and calm, and the air was full of chinch-bugs flying in all directions. Mr. Wells said that they had been flying in great hordes the preceding day. The fungus was no more abundant in the corn stubble (B) in the open field than at the time of his former visit, but the attack had increased in the shocks. Fifty-one chinch-bugs dead with muscardine were taken from a surface area of one square foot under a shock near the center of the field where the bugs were still con-centrated. Only a few were seen in the stubble. Grass along the lane and in the orchard between B and C was alive with bugs, but no dead were seen among them. The four-acre meadow between B and F was damaged to a slight degree, and all the meadows and pasture lands in the immediate vicinity of corn were more or less injured.

Several fungus-covered bugs were found half a mile east of the Wells farm, along a dead furrow, in a field of corn where the ground was quite damp and the corn considerably lodged. No Sporotrichum had been found in this field September 18. Evidence of the chinch-bug muscardine was also quite common in low, damp places in an adjacent field of corn. Five or six whitened bodies could commonly be seen under every fallen stalk. About the same condition existed in corn one-third of a mile south of the latter field. The fungus was quite common in a five-acre corn field west and north of C, on the opposite side of the road, belonging to Mr. Thomas Arington. It was easy to find a dozen or more dead bugs covered with Sporotrichum under almost any piece of fallen herbage, or under pumpkin vines, which were common throughout the field. About two and a half acres of the corn had been completely destroyed by chinch-bugs coming from wheat on the south side, leaving only an occa-sional stalk standing here and there. An hemipterous insect (*Nabis fusca*) imbedded in this fungus, was found on the ground under a cluster of grass by the roadside opposite the northwest corner of C.

The disease was also found in a twelve-acre field belonging to Mr. A. C. Rogers, one-half mile west of the Wells farm.   About half the corn in the field had been cut, the greater part of the remainder being flat on the ground.   The stalks were dwarfed, the leaves dry and brown; and the ears were little more than soft cobs, with an occasional imperfect grain attached to them.   Chinch-bugs, mostly adults and pupae, were very abundant behind leaves at bases of stalks, where the plants were somewhat green, and under every cluster of grass about the field.   In the open field, where the corn had been recently cut and where the ground was rather damp, eighty-three dead bugs imbedded in this fungus were counted under a single stalk, and whitened bodies could easily be found under any bit of rubbish or piece of herbage about the field. Two fungus-covered chinch-bugs were found in a field half a mile south, and traces of this fungus were found in all other fields examined in this vicinity at this time.   From the foregoing, it is clear that the white fungus was generally present in this neighborhood, having been found in all fields examined.

The final visit to this section was made by Mr. Johnson and myself October 10. The fungus-covered bugs were not so abundant in the open field (B) as formerly, although traces of the disease were still present on the ground under grass, , weeds, and rubbish of all kinds; but only an occasional dead bug was seen which bore a'fresh fungus growth.   As a rule, dirty whitish spots, scattered irregularly over the ground here and there, were all that remained of the older examples.

Very few live insects were seen in the corn stubble at this time.   An occasional cluster of foxtail-grass was thickly covered with adults and pupae, but in such places dead insects with a fresh fungus growth were very rarely seen.

The bugs were still accumulated in the shocks, especially along the south side of the field; perhaps because this was the last corn cut and was greener and better suited for food.   In such places fungus-covered bugs were quite common, many of them apparently just dead, as the white cottony growth was just appearing on their bodies.   The fungus attack, however, had not increased since our last visit, but on the other hand had perceptibly diminished.

Adults and pupae were quite numerous in grass along the lane north of B, but were less than half as abundant as nine days previous.   This reduction was probably due to the flight of the winged individuals, and not to any contagious disease, as not a single dead insect was found.   We did not examine the fields D and F, as Mr. Wells told us the condition of affairs was about the same as on our previous visit.

Sporotrichum was generally present in this region at this time, as shown by our finding fungus-covered bugs in corn shocks and in stubble two miles northeast of Farina, and in similar situations in corn on the farm of Mr. R. H. Smith, four and one-half miles east of the city.   The fungus growth on the bugs found at the latter place was fresh, but the disease had evidently been present in the field for some time past, as traces of old material were easily detected on the ground, in corn shocks, and in the stubble.   Mr. Smith said that no infected bugs had been distributed in the neighborhood of this field.   The occurrence of this disease seemed to be universally spontaneous at this time, as we found traces of it in all the surrounding counties.

Mr. Wells wrote November 20 that the disease was still present in corn (B), and reported having seen many fresh fungus-covered bugs in corn shocks while husking corn at that time.

With all the facts before us concerning the interesting occurrences upon this farm, we cannot say that our experiment was successful from the economic point of view, for the muscardine outbreak did not reach its maximum until after the corn had passed the growing season, and it was therefore of no practical use in protecting the crop from the ravages of the chinch-bug.   Neither can we say that it was certainly due to the artificial distribution of infected specimens, as the fungus was present here when the first lot of dead chinch-bugs was distributed.   (See No. 55.)   We must also note that an innumerable host of chinch-bugs remained in the corn in a perfectly vigorous condition during the entire dry period, which included the latter part of June, all of July, and the greater part of August; and that the muscardine fungus apparently disappeared with the advent of this dry weather, not attracting attention again until early in September, after the fall rains had set in, more than a month and a half from the time when the last infected bugs were distributed. Myriads of pupae and full grown chinch-bugs were present, indeed, in the very midst of the disease in September, and remained apparently healthy and vigorous until winter came on. The ratio of insects dead with muscardine to the live ones present in the field was insignificantly small to the last.

We are also in doubt whether the occurrence of the fungus on farms adjacent to that of Mr. Wells is to be attributed to its spread from his premises, especially as we found it early in October (from the 6th to the 13th) generally prevalent in the counties of Marion, Effingham, Clay, Jasper, Richland, Cumberland, Bond, Morgan, Sangamon, and Champaign. It seems quite possible, indeed, that its appearance on the Wells farm itself was due to the conditions that favored its general development at this time throughout the greater part of southern Illinois.

Finally, we have no really positive assurance that its growth and spread on Mr. Wells' farm was even hastened by his wholesale and persistent distribution of dead bugs, for the fungus was quite as abundant at the same time in far distant places, where only a few infected insects had been distributed (see No. 77), and in still others where no disease whatever had been artificially introduced (No. 73). It was also almost entirely absent in other distant localities, where large quantities of both cultivated material and infected insects had been scattered (Nos. 63, 64, 65, 66, and 67).

No. 58.   This is also a farmer's field infection experiment, the last of the series conducted on the Wells farm.   June 10 Mr. Wells placed part of the material taken from his contagion box (No. 56) on that date in the northwest corner of oats (C), where chinch-bugs were very numerous, having come from wheat on the opposite side of the road.   The material, about one-third of a quart, was scattered over the ground between the drill rows, covering a strip five or six feet in width by four or five rods in length (see heavy dotted lines on Plate I).   A second distribution, similar to the above, was made June 20, of several hundred chinch-bugs, dead with muscardine, taken from the same box (56).

No traces of the fungus were found in this field June 25, and Mr. Wells reported that no disease was present there July 17, when the oats were cut, although live chinch-bugs were everywhere abundant in the west half.   Mr. Johnson carefully searched the stubble and grass along the road September 19, but found no insects dead with disease; in fact, the bugs had entirely abandoned the field, except a few adults and pupae feeding on an occasional cluster of foxtail-grass.   The grass along the roadside was badly infested; but dead insects were very rarely seen.   The fungus was quite abundant in corn several rods to the northwest September 28, but the hemipterous insect found at that time, referred to in experiment 57, was the only fungus-covered bug seen in the immediate vicinity of the oat field.   The experiment was an utter failure so far as the destruction of chinch-bugs in the oats was concerned.

No. 59.   A farmer's contagion experiment conducted by Mr. James Smith on his farm near Farina.   About one dozen fungus-covered chinch-bugs from No. 54 were delivered to Mr. Smith June 7 by Mr. Marten.   June 8 a box was prepared according to our directions, in which a lot of live chinch-bugs collected from wheat were placed, together with the bugs dead with Sporotrichum.   The box was supplied with fresh food and live insects when needed.   About June 15 several dozen whitened bodies were removed and placed in wheat (No. 60).   The box was kept in good condition until about June 27, when it was abandoned by Mr. Smith, who at that time considered the contagious-disease method of "no account in checking chinch-bug ravages," and resorted to the furrow method for the arrest and destruction of the bugs (see No. 88).

No. 60. A farmer's field infection experiment made by Mr. James Smith, on his farm near Farina, with several dozen bugs dead with the fungus from No. 59, which were placed about June 15 in a wheat field where chinch-bugs were very numerous. June 19, field carefully examined by Mr. Marten. First search made in vicinity of place where infected bugs had been distributed. A few adults found here dead with muscardine, but others also throughout the field in about the same proportion. It must be borne in mind that this same fungus was generally present in this vicinity at this time, being more or less abundant in all fields visited. June 7, as noted under No. 55, it was found in wheat on the Wells farm and on other farms adjoining Mr. Smith's. It is quite probable, therefore, that the fungus of the white muscardine of the chinch-bug was locally present in the wheat where Mr. Smith first placed his material. July 11 Mr. Marten found no traces of the fungus in the wheat stubble or in corn adjoining. The disease practically disappeared on this place during the very dry weather of the latter part of June, the whole of July, and a part of August, and appeared again, in corn, late in September, when it was also generally present throughout this region. Mr. Johnson found several chinch-bugs dead with this fungus in an adjoining corn field on this farm September 18 and 28, but no traces were detected in the wheat stubble adjoining.

No. 61. June 15. A field infection experiment started by Mr. Marten in spring wheat on the University farm (see B, Plate II). A quantity of dead and fungus-bearing bugs, together with a few live ones, all from contagion box No. 54, were scattered on the surface of the ground at the bases of the wheat plants in the second drill row for a distance of several rods, represented at *c–a*, a place where live chinch-bugs were most numerous. Ground very dry, no rain having fallen since June 1, and then only .02 of an inch. Sky clear, temperature 91° Fah.* June 16, slight rain in the afternoon (.23 inch), temperature 82°. June 17, cloudy, rain (.2 inch), temperature 82°. June 18, cloudy, temperature 74°. Examined by Mr. Marten. Live bugs were numerous, and traces of original material still present, but no indication that disease was spreading. June 19, clear and warm, temperature 81°. June 20, light rain in afternoon (.1 inch), temperature 87°. June 21, cloudy, temperature 86°. Examined by Mr. E. B. Forbes. Not the slightest indication that the disease was spreading, only two adult chinch-bugs being found dead with fungus after a long-continued search, and these apparently a part of the original stock. Large numbers of young bugs in all stages, with a few adults intermingled, feeding freely at the bases of wheat plants, but no appearance whatever of the infection among them. June 22, slight rain (.05 inch) in the afternoon, temperature 88°. June 23, cloudy, temperature 90°. June 24, 25, and 26, considerable rain (1.18 inch), with average temperature 82°. June 27, clear, temperature 88°. Examined by Mr. Johnson and myself. Plot closely scrutinized throughout its entire length and breadth. Bugs very numerous, but none dead with fungus. Insects rapidly advancing into adjoining corn (C). This same day all the earth from Nos. 68, 69, and 70 was removed, together with several thousand live bugs and many dead ones (about 100 with the fungus and others without), and all scattered in the vicinity of the first place of distribution. June 28, 29, and 30, clear, with high temperature (average 89°). July 1, light rain in forenoon (.05 inch), temperature 83°. July 2, sky clear, temperature 85°. July 3, forenoon clear, afternoon cloudy, temperature 85°. Experiment carefully examined by Mr. E. B·Forbes. A few dead fungus-covered chinch-bugs found on ground in first drill row, but no others at any point in the field. Every wheat plant badly infested with bugs of all ages, mostly of the last two moults. The first three or four rows of corn adjoining blackened with pupae, or bugs of the moult just preceding, with an occasional adult, all feeding voraciously and apparently vigorous. No trace of disease among them. July 4, 5, 6, and 7, clear and warm, with no rain, average temperature 81°. Wheat

---

*Temperature taken each day at 2 P. M.

cut on the latter date.   Crop a complete failure; heads light and grains very small and shriveled; not gathered at all; burned over the following day.   Many chinch-bugs destroyed by the fire   Corn badly attacked.   Experiment a complete failure so far as the spread of the disease was concerned.

No. 62.   A second field infection experiment, made June 18 in the same strip of wheat on the University farm as No. 61, and under precisely similar conditions except that the material used consisted of about three thousand live chinch-bugs thoroughly exposed to infection in No. 54, and liberated by Mr. Marten at the point represented at *b*, Plate II.   Examined June 21 by Mr. E. B. Forbes, June 27 by myself and Mr. Johnson, and July 3 by Mr. E. B. Forbes again, but at no time were bugs dead with the white fungus found in sufficient numbers to indicate that the contagion had taken effect.   On the latter date, however, half a dozen fungus-covered bugs were found on the ground in the first three drill rows, where the original material had been scattered; otherwise no traces of the disease were seen, either in wheat or corn.   This experiment, like No. 61, considered a failure.

The five following (Nos. 63-67) are successive field infection experiments made on the farm of Mr. Samuel Bartley, one mile west of Edgewood, in southwest Effingham county, a locality particularly favorable to our purposes, since it was in the midst of one of the worst infested districts of southern Illinois.   These experiments were followed through the season by Messrs. Marten and Johnson, of the office force, assisted by Mr. Bartley on the ground.*

All the wheat fields on this farm, as well as those of the surrounding neighborhood, were closely examined by Mr. Marten June 19. Young chinch-bugs in all stages of development were everywhere abundant, covering the wheat in many places, and a few adults were seen.   In all these fields small numbers of chinch-bugs, both young and old, were found dead with the fungus of white muscardine, spontaneously occurring.   The ground was rather moist at the time from heavy rains of the 16th and 17th of June, the latest previous rain having fallen May 22.   The temperature since this latter date had been uniformly high, the daily record rarely falling below 90° Fah.†

No. 63.   June 19, several thousand chinch-bugs from No. 54, some still alive and others dead with the white fungus, were placed on the ground in a small wheat field on the Bartley farm where chinch-bugs were most abundant, the exact location being marked by two stakes driven into the ground.   Mr. Bartley kept a record of temperature and rainfall, and examined the field every third or fourth day.   June 20, 21, 22, and 23 were exceedingly hot, the average temperature reading for these days being 96¾° Fah.   No indications that the disease was spreading.   June 24, light rain. June 25 and 26, heavy rains and high temperature (average observation, 88°).   No dead seen.   On the other hand, the chinch-bug injury was increasing rapidly, and some wheat was wilting, with shriveled heads.   June 27, temperature 100°; wheat cut on 28th and 29th; dry and very hot, average midday temperature, 99°.

Examination by Mr. Marten June 29.   Several fungus-covered insects found in drill rows where material was placed, and a few others a short distance away, but as a rule their whitened bodies were no more numerous than before the infection was

---

*Mr. Bartley has been for several years a correspondent of the office with respect to the economic entomology of his district.   He is a man thoroughly competent by education, temperament, and experience to report upon such matters as were here entrusted to him.

†The temperature observations here reported were made by Mr. Bartley daily between eleven and twelve o'clock.

distributed. Myriads of living bugs in the stubble, but a general movement toward adjoining corn on one side and toward timothy on the other. June 30, a shower; temperature 98°. The average midday maximum for the whole month was 94°. July 1 to 10, very warm and dry, average midday temperature, 83½°.

Field examined again by Mr. Marten July 10. Wheat stubble very dry, yet containing many live bugs. The great majority, however, have gone into an adjacent corn field. No dead found in stubble, grass, or corn. The remainder of July hot and dry. Good rain the 18th, and slight shower the 19th. No dead found showing any traces of disease. Corn literally covered in many places. Light rain on 28th and 29th. No dead observed. Very warm on 30th and 31st, average temperature record 92°. Average midday reading for month, 88°. First three days of August very warm, with light rain the 3d, followed by a week of extremely hot weather. No dead seen.

Nos. 64 and 65. August 6, a second infection experiment was begun by Messrs. Marten and Johnson in two of Mr. Bartley's fields with material derived from two different sources. The first was about three inches square of a culture on corn meal saturated with beef broth from No. 2 (second remove from larva found April 17), and the second consisted of several hundred chinch-bugs dead with Sporotrichum from No. 68. The corn was badly dwarfed in both fields and literally alive with chinch-bugs. Ground damp, temperature 88°. No trace of the disease detected at this time in these fields or anywhere in the neighborhood.

The cultivated fungus (No. 64) was placed in corn about fifty rods from the spot where the first infection (No. 63) had been introduced into wheat. A row along a dead furrow was chosen, where the corn was stunted and literally covered with chinch-bugs. The culture material was cut into small fragments and dropped into the midst of the bugs behind every sheath of twenty-nine hills, the position being carefully marked by cutting away the tassels from the hills at either end.

The infected bugs (No. 65) were distributed in an adjoining corn field thirty-nine rods from the wheat and fifty rods from the preceding distribution (No. 63). Condition of corn about the same as in the foregoing (No. 64). Fourteen hills on the south end of the thirteenth row, counting from the west side, were treated and marked as above.

The weather continued dry and hot for the next four days, the average midday temperature being 97½°. August 11, heavy rain, accompanied by extremely hot weather, the thermometer registering 100°. Both fields critically examined by Mr. Bartley, but no dead bugs seen and only an occasional trace of the original material found. Examined again on the 17th, immediately after a slight shower. No fungus found. Corn in bad condition; ears shriveled and shrunken. The high temperature continued, the noonday average from the 12th to the 17th, inclusive, being 92°, and that for the month being 93°. Light rains on the 23d and 24th. Live bugs on the increase; much corn fallen down. No dead bugs, and no indication that fungus had spread. September 4, very heavy rain, followed by high temperature. Fields very carefully examined by Mr. Johnson. Very muddy, and much corn flat on the ground. In such places, especially, the bugs literally covered every stalk. Few adults seen, the great majority being of the first or second moult or pupae. Only three bugs dead with this fungus found after a long-continued search, and these under fallen corn, on the ground, a considerable distance from the place where the infection material was placed. Only an occasional stalk of the hills originally treated remained erect, the rest being dead and flat on the ground.

No. 66. Conditions being now especially favorable to success, several hundred spore-covered bugs from No. 68 were distributed in this field on the Bartley farm by Mr. Johnson September 4, behind leaf sheaths, and on the ground under the fallen stalks of twenty-three hills which were black with bugs. The location, which was about twenty-five rods from that of experiment 65, was marked as before. September

7, no indication that the disease is spreading; a few traces of the original material present. Bugs everywhere abundant, and seemingly healthy. September 10, ground still damp, and corn covered with bugs, but none dead with fungus. September 18, similar report.

No. 67. September 7. A second lot of cultivated fungus introduced on this date, material from No. 2 having been sent to Mr. Bartley, who placed it, according to directions, in a one-acre strip of late corn near his house, about eighty rods from the other experimental fields. The chinch-bugs had accumulated in great numbers in this late-planted patch. Seventeen hills of the seventh row, in the southwest corner, were thoroughly treated, as in No. 66. The ground was quite damp, and September 12, a heavy rain fell, followed by extremely hot weather. No dead bugs seen at this last date, but traces of infection material present behind leaf sheaths and on the ground among the bugs. September 16, light rain; corn beginning to wilt from chinch-bug attack; not a single dead insect seen; a little culture material still present.

September 18 Mr. Johnson carefully examined the Bartley farm, and found only a single fungus-covered insect in the last-mentioned field. No traces of the cultivated material. Corn about all dead, a large proportion of it being flat on the ground. Corn in adjoining fields all dead in the immediate vicinity of the places where infection experiments were made, and only six bugs dead with the fungus found after long-continued search. The drooping tassels, and the dirty brown, ragged leaves hanging close to the short dwarfed stalks of the remainder, gave to the whole neighborhood an aspect of desolation. In some places the bugs still blackened the stalks but there was a general movement toward an adjacent meadow, in which considerable damage had already been done. Observations made at later visits, did not disclose any fungus-covered insects on this farm, and we must therefore class this series of experiments, Nos. 63-67, as absolute failures.

The four following (68–71) are parallel laboratory contagion experiments, conducted at Natural History Hall in large covered wooden boxes, each six feet long by three wide and six inches deep, supplied with a layer of dirt half an inch deep, freed from leaves and rubbish and thoroughly moistened, the bottom of each being finally covered with a layer of fresh green oats, or the stalks and leaves of corn, for food. Each box was supplied with fresh food and live insects as circumstances required, usually every second or third day. The period of operations extended from June 22 to September 20, during which time a sufficient quantity of infected insects was taken out to supply nearly two thousand applicants throughout southern, central, and northern Illinois, as well as large quantities used in our own experiments.

No. 68. June 22 a part of the contents of 54 were transferred to this box, together with a large quantity of live chinch-bugs from the field. Everything was thoroughly moistened and the box closed by means of a tight-fitting cover, screwed down, the cracks being afterwards pasted up with narrow strips of paper. June 25, found no very considerable development of the fungus. Large numbers of bugs dead which showed no appearance of the infection. Ten of these crushed in water on a slide and examined microscopically contained no fungus mycelium. One lot taken from earth at this time showed no traces of the disease; while several others taken from corn leaves were everywhere penetrated with mycelial threads of some kind of fungus. Dead insects removed and placed on damp sand for further observation. Spoiled food removed, and fresh oats introduced. June 26, quite a number of fungus-covered bugs picked out from food and from surface of dirt; fresh feed supplied and more live insects. June 27, box in bad condition; all the earth, food, and insects removed.

Box cleaned and re-stocked with fresh dirt, food, and live insects. Several of the whitened bodies taken out were returned to the box. The old material, dirt, food, insects and all, scattered in wheat (Experiment 61). June 28, food renewed and live chinch-bugs introduced. June 29 and 30, fresh food supplied.

July 1, box in very bad condition; considerable mould on the dirt, and many bugs covered with Aspergillus. Comparatively few covered with Sporotrichum. All the material removed, box washed out with soap and water, thoroughly burned over by means of a Bunsen burner, and wet with alcohol and burned out the second time. The bottom covered with moist sand about half an inch deep, fresh food introduced, re-stocked with live insects, and infected with fungus-covered bugs collected by Mr. Marten from fields at Odin June 22, and at Shattuc June 23. July 2, no appearance of the fungus. July 3, in good condition, but no fungus developing. Fresh food and live insects from reception box introduced. July 4, very clean and free from mould; very little fungus present; growth not rapid; food changed. July 5, fresh food and live insects added. Anguillulids noticed in abundance in the earth. July 6, a few chinch-bugs with Sporotrichum. Fresh food and live bugs introduced. July 8, no bugs with fungus seen. Box in clean condition. Four cicadas, dead and well covered with this same fungus, collected at Mahomet by Mr. E. B. Forbes, were broken up and the fragments distributed along the sides and corners of the box where live chinch-bugs were most numerous. July 10, boxes overhauled and cleaned and fresh food and more bugs introduced from reception box   A small number of insects dead with the fungus seen. July 11, very little Sporotrichum seen. Box in good condition. Fresh food added. July 12, fresh food introduced. Box in fair condition. July 13, about as yesterday. July 14, fungus scant; fresh food. July 16, three or four bugs with fungus seen; fresh food and live bugs introduced, July 17, no fungus found. Part of the food changed. July 18, very little fungus seen. Fresh food and more bugs introduced. The amount of water used in box increased.

July 19, about half a dozen bugs dead with the fungus. A fourth infection, coming from No. 11, a culture on agar, was introduced at this time. July 20, but little fungus seen. Slight mould on earth. Spoiled food removed. July 21, box overhauled and fresh food introduced. Fungus about as yesterday. July 23, five or six bugs with Sporotrichum seen. The box contains less mould than any of the others (69, 70, and 71) and fewest fungus-covered insects. More live chinch-bugs from reception box added. July 25, overhauled and food renewed. Only about half a dozen fungus-covered bugs seen. Many anguillulids were found quite abundant on bugs dead with and without the fungus, also on the culture medium. A piece of this material upon which both· Sporotrichum and Aspergillus were growing was washed and found to contain a considerable number of anguillulids both dead and alive, confined principally to the surface, or slightly imbedded in the softer and somewhat decomposed spots. July 26, Sporotrichum very scarce. July 27, transferred contents, · except sand, dead insects, and a few live bugs, to No. 69. Very little fungus seen. July 30, fungus greatly increased. 165 insects dead with this disease removed. Sand and interior of box thoroughly wet. July 31, quite a number of dead bugs have a fresh fungus growth appearing. A few mites seen.

August 1, seventy insects dead with Sporotrichum removed. Large numbers of chinch-bug eggs found on the sand and on the older corn leaves. Young chinch-bugs, recently hatched, had been noted here for several days. A considerable number of young insects also present. All the sand removed, box scraped and brushed, allowed to stand until nearly dry, then wet with alcohol and burned out   A layer of fresh earth from half to three-quarters of an inch deep placed in bottom and cultivated fungus on agar re-introduced. The garden soil was taken from five to twelve inches below the surface, and was free from rubbish. August 2, all the live chinch-bugs in No. 70 placed in this box. August 4, large number of dead insects, but no fungus seen. Box in good condition. Fresh food introduced. August 6, many bugs dead with good growth of fungus and many others without it.   Both Aspergillus and mites in small numbers on culture material.   Food renewed.   August 9, a large number of dead insects developing Sporotrichum freely.   A very few mites and some Aspergillus found on the culture material, much of which was taken out of the box.   Food renewed.   Several anguillulids found on dead insects in the dirt.   August 11, bugs dead with the white fungus on the increase.   Aspergillus also present in considerable quantities.   A few mites seen on culture material.   Anguillulids very abundant, filling the body cavities of insects dead for a considerable length of time.   Chinch-bug eggs very abundant on the culture material, some apparently about ready to hatch.   August 15, condition of box about the same as when last examined.   Sporotrichum still present in considerable quantity.   August 18, amount of fungus about the same.   Fresh food supplied, box thoroughly moistened.   August 22, many adults and a few young dead with Sporotrichum, and several covered with Aspergillus.   Still another greenish fungus was present on several bugs.   Eggs hatching by thousands in this box.   Mites increasing, and anguillulids still present in small numbers.   Fresh food supplied.   August 25, a considerable number of insects dead with Sporotrichum.   Aspergillus seen in small quantities.   Live bugs, mostly adults, quite numerous.   A few young seen.   Fresh food supplied.

September 3.   Sporotrichum still quite abundant.   Several hundred whitened bodies picked from the surface of the earth and used for experimental purposes by Mr. Johnson at Edgewood and Odin September 4 and 5 (see Nos. 66, 73, and 74).   A few live bugs still present.   Fresh food introduced.   September 16, box overhauled and fresh food added.   About as many fungus-covered bugs as on last date.   About fifty dead with muscardine removed and used for field infection experiment at Odin September 19 (see No. 73).   Fresh food introduced.   September 20, final overhauling.   Several hundred whitened bodies picked out, which were distributed by Mr. Johnson to farmers in the south-central part of the state early in October.   Only an occasional live insect seen.   Box discontinued.

No 69.   June 23, a second contagion box, precisely similar to No. 68, stocked with a large quantity of live chinch-bugs received from correspondents.   The infection introduced was from No. 54.   Examined June 26; picked out a few fungus-covered bugs and added fresh food and chinch-bugs.   June 27, all the material, dirt, bugs, and food removed and scattered in wheat (61), except a few whitened bodies reserved for re stocking the box.   Box thoroughly cleaned.   June 28, fresh food supplied and a considerable number of live bugs introduced with the fungus-covered bugs removed yesterday.   June 29, fresh earth introduced, and cultivated spores on agar, from No. 11, added.

July 1, very bad condition; earth covered with mould and many dead insects enveloped in Aspergillus.   All the contents removed and box thoroughly disinfected.   Re-stocked with fresh earth, fresh food, and live bugs.   Infected with a number of fungus-covered insects collected by Mr. Marten from fields in the vicinity of Odin (June 22) and Shattuc (June 23).   July 2, fresh food added.   July 3, food and an additional lot of bugs introduced.   Box in good condition, but no muscardine fungus.   July 4, fair condition, but little Sporotrichum.   Food changed.   July 5, about the same as yesterday, except that many anguillulids were seen.   Fresh food and more live bugs added.   July 6, a little Sporotrichum present, but not so abundant as in No. 70.   Mould still growing on earth.   Fresh food and live insects introduced.   July 8, a few bugs dead with the fungus removed, but the yield was very poor.   Two cicadas dead with this same white fungus, from same source as those used in No. 68, were broken up and distributed along the sides, and cultivated fungus on agar from experiment 11, was scattered over the surface of dirt through the middle of the box.   July 10, a little Sporotrichum present.   Box cleaned, fresh food supplied, and live bugs from reception box introduced.   July 11, box overhauled and very little fungus seen.   Fresh food introduced.   July 12, in fair condition; fresh food added.

Anguillulids present in small numbers. Many bugs dead, but show no traces of the fungus on their bodies. Attempts to develop Sporotrichum on these bugs by placing them on damp sand proved failures. July 13, condition about the same as yesterday. July 14, Sporotrichum scant. Box put in good condition. Fresh food supplied. July 16, only three or four fungus-covered bugs seen. Moulds not so bad as formerly. Fresh food and live insects from reception boxes added. July 17, half a dozen. ungus-cov ered bugs removed. Part of the food changed. Box in fair condition fJuly 18, very little fungus seen. Fresh food and more live bugs introduced, and the amount of water used in moistening the box increased. July 19, four or five bugs with Sporotrichum. Enough live insects taken out to fill 187 pill boxes, which contained also cultivated Sporotrichum (from No. 2) for distribution. Box in good condition. July 20, but few fungus-covered bugs seen. Spoiled food removed. Little mould present. July 21, fungus about as yesterday. Fresh food introduced. July 23, Sporotrichum not very plentiful. A few bugs dead with the disease found on and in the earth. One hundred and forty-four pill boxes filled with live insects and cultivated material (from No. 2). More live bugs added. July 24, small number of bugs with Sporotrichum seen. July 25, box overhauled and food changed. July 26, white fungus very scant. July 27, very little Sporotrichum seen. No anguillulids found in earth or on dead insects. The contents of No. 68, except sand, dead bugs, and a few live insects, were transferred to this box. July 28, few fungus-covered bugs seen. Food changed. July 30, Sporotrichum increasing, but not abundant. Aspergillus also increasing. Anguillulids present in small numbers. Mites numerous on earth, but not abundant on dead or live insects. Fresh food introduced. July 31, about fifty bugs dead with Sporotrichum. Many insects with Aspergillus also. Moderate number of mites on dead chinch-bugs and decayed vegetable matter. Anguillulids present on dead bugs and dead coccinellid. Fresh food supplied—about one-half the quantity heretofore used.

August 3, final overhauling and cleaning up. Bugs dead with Sporotrichum estimated at about two thousand; many of them pupae, mainly on the surface of the dirt. Most of the hidden ones were under clods and loose lumps of earth. Quite a number of chinch-bugs were found dead in copulation, in most cases both sexes being infected with the white fungus. Occasionally, however, only one of a pair was dead or visibly diseased, the other being still alive, but unable to free itself. Two instances were noted where one of each pair was dead and covered with Aspergillus, while the other showed no growth of any kind. Mites still present, but considerably less numerous than before. Live chinch-bugs were not abundant, those remaining being mostly adults, with very few pupae, young of the first and second moult, and numbers of eggs. Box discontinued.

No. 70. June 23, a third box, similar to No. 68 and 69, was stocked with food and large numbers of chinch-bugs received from correspondents, but not infected. June 27, thoroughly overhauled; earth supplied; new food and more live bugs introduced. Infected with fungus-covered bugs from experiments 68 and 69. June 28, food renewed and more live bugs added. June 29 and 30, fresh food introduced.

July 1, mould on surface of earth, box in very bad condition. Aspergillus present on several dead bugs. Comparatively little Sporotrichum. Box thoroughly cleaned and disinfected; earth renewed, fresh food added, and live insects introduced. July 2, no fungus seen. July 3, in good condition, but no Sporotrichum seen. Fresh food and more bugs introduced. July 4, in fair condition. A few insects with fresh fungus growth on their bodies. Food changed. July 5, condition about as yesterday, except that anguillulids were noticed quite abundant in earth. Food changed and more live insects added. July 6, white fungus more abundant than in experiment 69. Mould on dirt still spreading. Fresh food and more bugs added. July 8, a few chinch-bugs dead with white fungus removed. Earth stirred to destroy mould. Box in fair condition. A second infection introduced. Two cicadas from same source

as those used in Nos. 68 and 69, were broken up and distributed over the surface of
the dirt, along the middle of the box, and a quantity of cultivated material from No.
2 was distributed around the sides.   July 10, a few fungus-covered bugs removed.
Dirt contains much mould. Aspergillus less abundant than previously. Box cleaned
up and fresh food introduced. July 11, very little fungus seen. Fresh food added.
July 12, little Sporotrichum present. Anguillulids seen in dirt. Many bugs dying,
but without developing Sporotrichum. Fresh food introduced. July 13, about same
condition as yesterday. July 14, Sporotrichum scant, but more abundant than in Nos.
68 and 69. Fresh food added. July 16, thirty-six fungus-covered bugs picked from
surface of dirt, and afterwards returned. Moulds not so bad as formerly. Fresh food
and more bugs from reception box added. July 17, no fungus of any consequence
present. Spoiled food removed. July 18, very little Sporotrichum seen, but more
abundant than in other boxes (68, 69, and 71). Food changed and amount of water
used in the box increased. July 19, fungus more abundant than in 68 and 69. Box
in good condition. July 20, very few fungus-covered bugs seen. Removed enough
live bugs for three hundred pill boxes, with cultivated fungus (from No. 2), for shipment. A little mould present on earth. Old food removed. July 21, fungus about
as abundant as yesterday. Box in good condition. Fresh food introduced. July 23,
few bugs with fungus seen. One hundred and forty-four pill boxes filled with live
insects and cultivated fungus from same source as above, and prepared for shipment.
Large number of live bugs from reception box introduced. July 24, fungus-covered
bugs more abundant, but not numerous. Anguillulids present on dead chinch-bugs.
July 25, overhauled and food renewed. July 26, Sporotrichum very scant; the
living bugs from experiment 71 transferred to this box. Anguillulids found in
earth and on corn leaves in small numbers. July 28, few bugs with Sporotrichum.
seen. Fresh food introduced. July 30, number of bugs with fungus not great.
Aspergillus on the increase. Mites abundant on the earth. Anguillulids present on
dead pupae and adults. The worms were present on a dead pupa which contained an
apparent mycelial growth of the white fungus. Two adults, dead and badly decomposed, contained them in considerable numbers. July 31, about fifty insects dead
with the white fungus seen, but growth very poor, in most cases being quite incon-
spicuous. Mites very abundant on the corn, especially where the sap was exuding.
A few were seen on dead chinch-bugs behind corn leaves. Anguillulids were also
seen on dead bugs in similar situations. Old food removed, and about half the
quantity of fresh food usually introduced was added.

August 2, two hundred and eighty-one fungus-covered insects removed. Among
hem a pair copulating, the female being dead, with a short mycelial growth about the
thoracic region, while the male was still alive, quite active, and showed no signs of
distress or irritation. Chinch-bug eggs and young were abundant. Ten blow-fly
larvae were seen. Mites very abundant, sometimes accumulating in great numbers
on the leaves. 'Box finally overhauled; everything removed and transferred to No. 68.

No. 71 June 26, the fourth large box was supplied with earth, and stocked with live
chinch-bugs and fresh food, but no infection. June 27, fresh food introduced. June
28, food renewed and a lot of live bugs added. Infected with cultivated spores on
corn meal and beef broth (from No. 2). June 29, fresh food introduced. July 1, in
fair condition. Cleaned up. Fresh food and more live insects added. July 2, very
little Sporotrichum seen. A little mould on the earth. Fresh food introduced. July
3, enough fungus-covered bugs removed to make up forty-nine boxes for correspondents. Little mould on earth. More live bugs added. July 4, fungus quite
abundant, the whitened bodies of the insects being most abundant just beneath the
surface of the earth. Material for thirty-eight packages removed. Earth in box
stirred up to destroy moulds. Fair condition. Food changed. July 5, condition
about as yesterday, except that a few anguillulids were seen in earth. July 6, a few
fungus-covered bugs present. Moulds still developing on earth. Fresh food and

live insects added. July 8, very little Sporotrichum seen. Box in very bad condition. Aspergillus quite abundant on dead bugs, and moulds common on the earth. Blow-fly larvae very numerous. Food changed and live bugs introduced. July 10, only slight traces of muscardine. Aspergillus seems less abundant. Moulds still present. Fresh food added and more chinch-bugs from reception box introduced. July 11, few bugs dead with Sporotrichum. Box in fair condition. Fresh food and more live insects added. July 12, in fair condition. Anguillulids present on earth. Many bugs dying without developing any sort of fungus growth. Fresh food introduced. July 13, condition about the same as yesterday. Anguillulids present in small numbers. July 14, Sporotrichum very scant. Food changed. July 16, only two insects dead with muscardine seen. Moulds on earth not so bad as formerly. Fresh food and an additional lot of bugs from reception box introduced. July 17, no Sporotrichum of any consequence seen. Fresh food added. July 18, white fungus very scant. Fresh food added. Amount of water used in moistening box increased. July 19, a greater number of insects dead with muscardine than were seen in experiments 68 and 69 on this date, the number dead, however, being about equal to those in No. 70. Box in good condition. July 20, very few fungus-covered bugs seen. Bad food removed. Slight traces of mould on earth. July 21, amount of fungus about as yesterday. One hundred and thirty-two pill boxes filled with live bugs and cultivated material (from No. 2), and prepared for shipment. Box thoroughly overhauled and fresh food introduced. July 23, more bugs from reception box introduced. July 24, bugs with Sporotrichum few in number. Anguillulids quite plentiful on dead insects. July 25, box overhauled and food renewed. July 26, white fungus very scant. A quantity of living bugs transferred from this box to No. 70. Anguillulids found quite abundant on the earth. July 30, final overhauling. No Sporotrichum found. Many dead insects covered with Aspergillus. Mites very abundant on earth, and their eggs found on a dead chinch-bug enveloped in a growth of Aspergillus. Box discontinued.

No. 72. An experiment begun June 1 to test the direct effect of moisture upon chinch-bugs in confinement. A number of specimens from Tonti, in southern Illinois, which had not been exposed to fungous infection, were placed in a Riley breeding-cage, the top of which was afterwards covered with glass to prevent evaporation. This cage sat in a metal pan, the bottom of the cage being filled with earth. Water was poured into the pan outside of the cage, and kept standing there continuously to insure the saturation of the earth and the air inside the cage.

June 2, water stood in drops on the sides of the cage and on corn and grass within. June 4, no losses among the chinch-bugs. June 9, many of the bugs dead, but with no trace of fungus growth. June 13, adults now all dead, but with no appearance of disease; young still in good condition; earth saturated and moisture standing in drops on plants and all over the inside of the breeding cage. June 15, only a very few young bugs left alive, none having shown fungous disease at any time. Between this and July 3, those remaining died; and at the latter date, when the cage was overhauled, no growth of a fungus parasite had appeared on these specimens.

The two following (Nos. 73 and 74) are successive field infection experiments conducted by Mr. Johnson near Odin, in Marion county, a district especially favorable for experimental work, each being followed up by repeated visits, from August 7 to October 6.

No. 73. August 7, several hundred fungus-covered bugs from No. 68, were placed behind leaf sheaths and on the ground in a field of late corn on the farm of Mr. Thomas Ferguson, Sr., one mile north of Odin. The spot chosen was especially favorable to the growth of the white fungus, being along a dead furrow in a low, damp place where the corn was much dwarfed and completely covered with chinch-bugs, which had come largely from an adjoining wheat field. No fungus-covered insects were seen here at the time.

Every hill on the west end of the twenty-third row, counting from the south side, was treated with spores of muscardine for a distance of thirty paces   The tassels were cut, marking the east and west boundaries. Chinch-bugs, both young and old, were very abundant; adults were copulating; and young just past the first moult reddened the stalks.

This field was examined by Mr. Johnson September 5. Very heavy rain had fallen the day before, and the field was very muddy. No traces of disease were found, except a single fungus-covered insect behind a leaf. The corn was nearly destroyed by the chinch-bugs, and flat on the ground in many places, especially in that part of the field where the infection material had been distributed. Only five or six stalks of the hills treated remained standing. Stalks flat on the ground and not dead were literally covered with live bugs. Other parts of the field, where the corn was most vigorous, and where the bugs were least abundant on our former visit, were now overrun with the chinch-bug hosts; in one place about two-thirds of a pint were collected from two hills, by jarring and shaking them over a cloth spread on the ground. The insects were very active, and apparently in a healthy condition.

At this date (September 5) Mr. Johnson placed a second lot of infected bugs from experiment 68 in this field, where the conditions were especially favorable. The ground was very wet and the corn much lodged and covered with chinch-bugs. About one hundred bugs, dead with the white fungus, were scattered behind corn leaves, about forty rods from the place where the first infection material was introduced.

September 19, Mr. Johnson carefully examined this field again. Corn had been cut, where it was worth saving at all, and shocked. The bugs were thickly concentrated in the shocks, but no traces of muscardine were seen among them. The insects were still very abundant in the stubble. The corn in many places was as flat on the ground as if a roller had gone over it, and in such places every stalk was blackened with bugs. A third lot of chinch bugs, about fifty in all, dead with Sporotrichum from the same source as the others (No. 68), were now scattered over the ground under the fallen corn, at one place about ten rods from the row where the second lot of bugs had been placed. The greater part of the corn on the ground was brown and dead, and there was a general movement of the chinch-bug hordes into late corn in an adjoining field on the west, and into a meadow touching the southwest corner. Considerable damage had been done by the insects to the grass in the latter field, and corn in the former was suffering severely from their attacks. Not a single diseased bug was found in any of these fields at this time.

Examined by Mr. Johnson September 26. Very few chinch-bugs in the stubble. One bug found imbedded in the white fungus. Many insects in the meadow. The late corn in the field adjoining the stubble badly damaged, the attack apparently increasing. No traces of disease seen in this field. Bugs seemingly vigorous and healthy.

The final visit of the season to this locality was made by Mr. Johnson October 6. At this time the late corn in the field last mentioned in the preceding paragraph had been cut and shocked about a week, and the ground planted in wheat. The bugs had accumulated in enormous numbers in the shocks. Half a dozen fungus-covered insects were taken from shocks in this field at various places, and several were taken from grass in the adjoining meadow, where the chinch-bug attack was still spreading. A dozen or more whitened bodies were taken from shocks and under fallen corn in the field where the infection had been distributed, but the fungus was nowhere abundant, and long-continued search was required to find even a single specimen. A few live chinch-bugs, mostly adults and pupae, were still present in the stubble, and bugs in the same stages were quite abundant in the shocks. Corn in both these fields was a complete failure, and was saved for fodder only. Chinch-bugs dead with the white fungus were found in all fields examined in this county at this time, but only in two places (Nos. 57 and 76) were they at all common, and even here the live bugs out-numbered the dead many thousand times. This experiment for the introduction and increase of the fungus by artificial means was to all appearance a complete failure.

No. 74. August 7, the second of this series of experiments was begun in a twenty-acre corn field (B, Plate IV) owned by Mr. Silver, of Cincinnati, Ohio, and planted by Mr. Frank Robinson, of Odin. This farm is about a mile and a half north of Odin, between the Ferguson farm on the south (No. 73) and the Hurd and Robinson farms on the north (Plate IV). The field was very dry and dusty; the corn thin, short, and very poor. Chinch-bugs just past the first moult covered nearly every stalk. A few adults were seen copulating. No traces of the fungus disease were seen.

A row of dwarfed, sickly looking corn along a dead furrow about the center of the field, thickly covered with bugs, was selected as a suitable place for the introduction of the infection. A piece of culture material from No. 2, about two inches wide by four inches long, containing a profuse growth of Sporotrichum, was cut into small fragments and placed among the bugs behind the leaves of every hill, for a distance of fifty paces, the spot being marked by clipping the tassels from the hills at either end.

Examined August 18 by Mr. Johnson. No fungus-covered insects found. Some of the culture material still present. Corn in very bad condition and flat on the ground in many places. Chinch-bugs cover the corn throughout the field. Foxtail and other grasses literally alive with the insects. Weather very dry and hot.

Second examination September 5, immediately after very heavy rain. Field very muddy. No traces of the infection material visible, and no bugs dead with the fungus seen. Corn crop an utter failure. Chinch-bugs seem to be increasing in numbers. Selected another row along a dead furrow, about five rods from the first, and distributed about one hundred fungus-covered bugs from experiment 68 behind the leaves for a distance of seventy-three paces, and marked as before.

September 19, not a single insect dead with the white fungus found in this field after an hour's diligent search. Corn about all dead. Chinch-bugs still very abundant, but many going into meadows on the north and east. Few adults seen flying.

September 26, about as before, except that the corn is all dead and chinch-bugs less numerous. No dead seen, and not a trace of the fungus found. Chinch-bugs very abundant in the adjoining meadows, and considerable grass killed.

The final visit for the season was made October 6. About one-fourth of the corn had been cut and shocked and saved for fodder; the other three-fourths was flat on the ground, dead and brown. Chinch-bugs, mostly adults and pupae, were quite numerous in the shocks, and on the fallen corn throughout the field. Half a dozen bugs dead with Sporotrichum were picked up on the ground under a corn shock about two rods from the place where the last lot of infection material had been introduced, and a few others were found at various points in the field, under shocks and fallen corn, but this was all. The presence of the fungus on chinch-bugs in this field at this time did not seem to have any connection with that distributed August 7 and September 5, since the disease was found more or less prevalent in all fields visited in this neighborhood at this time. As the white muscardine did not spread from the centers where the infection was introduced, and as the chinch-bug hosts continually increased, remaining in a vigorous and perfectly healthy condition in its very midst, the experiment is regarded as a complete failure.

This description should be read, however, in connection with that given under No. 76, relating to a considerable outbreak of chinch-bug muscardine on a farm immediately adjoining this upon the north.

No. 75. This is a farmer's field infection experiment made by Mr. Frank H. Robinson on his farm (the right hand third of Plate IV.), about two miles north of Odin. August 7, several pieces of culture material containing a profuse growth of the white fungus, from No. 2, were placed in Mr. Robinson's hands by Messrs. Marten and Johnson, and directions given for its distribution. August 8, two small plats of corn, represented on Plate IV., at C and D, thickly covered with chinch-bugs, were chosen and the infection material scattered along the north side next the orchard, behind the leaves and on the ground at the bases of the stalks where the insects were most abundant.

September 5, examined by Mr. Johnson. Corn all dead in plat represented at D. Chinch-bugs still very abundant on the dead brown stalks and leaves. No traces of the infection material found, and only two bugs dead with this fungus were seen; and these were taken from under a clod on the ground, which was very wet from recent heavy rains. About the same condition was noted in plat C, except that a few traces of the old culture material still remained on the ground at the bases of a few stalks, and that no fungus-covered bugs were seen. Chinch-bugs were present in small numbers in all the meadows bordering the field containing this corn. In a twenty-two-acre field north of the house, represented at E, the corn was completely destroyed by their attacks, and nothing green remained. The bugs were everywhere-abundant, and almost completely covered the ground in many places, but there were no traces of fungous disease. Grass in the meadow adjoining on the south was slightly damaged.

September 19, a few live bugs, mostly adults and pupae, were seen in the stubble at C and D, but no traces of Sporotrichum were found. The insects were still present in the surrounding meadows. The corn ground (E) had been planted in wheat, and only a few bugs remained in the field. The attack had greatly increased in the meadow to the south, and considerable grass had been killed. No traces of the white fungus seen on this place at this time. September 26 and October 6, similar report made by Mr. Johnson, except that a few fungus-covered bugs were found in the meadow south of E, on the latter date. The experiment is classed with 73 and 74 as a total failure.

This description should be read, however, in connection with that given under 76, relating to a considerable outbreak of chinch-bug muscardine on a farm immediately adjoining this on the west.

No. 76. This is the second of the three exceptional cases of spontaneous muscardine, referred to above. It appeared on the farm of Mr. Silas Hurd, about one and a half miles north of Odin, in the corn field marked A, Plate IV. This field of forty-nine acres was planted early and grew rapidly for a time, but then came practically to a stand on account of incessant chinch-bug attack and the drought which prevailed throughout that region during the latter part of the summer.

The corn was cut and shocked while yet in roasting ears, early in September, in order to save the fodder, and the field was deeply harrowed and planted to wheat. No Sporotrichum had been distributed in this field at any time; but unsuccessful attempts to infect fields had been made, as described under Nos. 74 and 75, on the Robinson and Silver farms at distances of a little more than a quarter of a mile away. (See B, C, and D, Plate IV.)

Mr. Johnson examined this field October 6, and found the white fungus quite abundant in all the shocks. One hundred and fifty-two dead chinch-bugs imbedded in it were collected in a few minutes from a single shock at *a*, and every shock examined in the southern and western parts of the field contained fungus-covered bugs in considerable numbers. At *b*, in the northeastern part, it was an easy task to collect several hundred whitened bodies in and under every shock. This part of the field was quite low, and chinch-bugs, mostly adults and pupae, had accumulated in the shocks in enormous numbers. This was perhaps due to the fact that all green vegetation in the field had been destroyed by the harrow and cultivator, and that the bugs were obliged to congregate in the shocks or to leave the field in search of other food.

Only an occasional fungus-covered insect was found in a field of sweet corn (F) which had been completely ruined by chinch-bugs, and which was not cut at all. Adults and pupae were still present there, but not very abundant. Bugs were seen in small numbers in all the green lands surrounding the corn (A); but no traces of the fungous disease were found, with the exception of three or four dead insects taken from grass in the meadow next the road east of A. A few others imbedded in Sporotrichum were found in corn (B), as noted in experiment 74, and several were taken from grass in the meadow west of corn field E, as stated in No. 75.

The fact that this fungous disease was almost totally absent in fields surrounding A, would seem to indicate that it must have been fostered in the latter field by especially favorable conditions. These seem to have been (1) the early cutting and shocking of the corn while it was still green; (2) the destruction of all food throughout the open field, such as grasses of various kinds and the green stubble itself, by harrowing and cultivating; (3) the consequent concentration of the chinch-bug hordes in the shocks; (4) the heavy rains which fell about September 16 and 17, wetting the shocks and thoroughly drenching the chinch-bugs; and (5) the retention of the chinch-bugs in the shocks at a time when their food supply was short, and when the moisture was also sufficient for the germination of any spores of the white fungus that may have been present.

Mr. Hurd wrote the office, November 20, that the disease was still present in these shocks, and that he had collected an abundant supply for use next spring.

No. 77. This is a farmer's field contagion experiment made by Mr. George W. Heth on his farm (see Pl. III.) about five miles west of Edgewood, in West township, in the extreme southwest corner of Effingham county. It is the last of the three exceptional cases of the development of white muscardine in the field, to which reference has already been made.

The material used in this experiment was derived from two original sources; the first, a small number of chinch-bugs dead with Sporotrichum received by Mr. Heth May 15, from Chancellor Snow, of the University of Kansas; the second, similar material sent him early in June from my office, from No. 54 of this series. The fungus-covered specimens received from Kansas were scattered directly (May 15) among chinch-bugs in wheat (see A, Pl. III.) at about the center of the field. The bugs were very abundant at the time, practically covering the wheat everywhere throughout the field. A slight rain fell May 17, followed by heavy storms on each succeeding day until the 20th. The fungus did not seem to spread, and the chinch-bug attack became daily more intense. Dry weather followed until the middle of June, when, according to Mr. Heth's observations, no traces of disease could be found in the field. In the meantime the second lot of specimens above mentioned, derived from No. 54, had been placed by Mr. Heth in a large pasteboard box stocked June 10 in the usual manner for infection purposes, but, according to his somewhat vague statement, without very successful results. Several days afterwards Mr. Heth began to distribute this material in his wheat. "I removed some of the insects from the box," he says, "part of which were white with fungus, every second day, and scattered them over the ground near the center of the wheat field A, where the bugs were extremely thick. These distributions were made in the evening and were kept up for about a week, but the soil was very dry." Rains fell on the 16th and 17th of June, but no traces of muscardine could be found on the 18th, or at the time when the wheat was cut, June 25. The ground was, in fact, in many places almost black with living chinch-bugs at that time, and there was a general movement to the north and east into corn fields adjoining (marked B and C on Pl. III.), the greater number going, however, into field C, where the corn was presently almost completely destroyed. The destructive horde passed thence successively into fields D, E, F, and G, where they continued their ravages until late in the fall.

On September 28, while examining the grass in the orchard F, at the point *a*, Mr. Johnson discovered many dead chinch-bugs on the ground under the stools of grass thickly enveloped in a dense, fresh growth of Sporotrichum. Half a dozen whitened bodies were taken from under a single stool at this point; but for every dead bug collected, several hundred live ones were seen. The fungus-covered insects were about equally distributed throughout F and G, along a line north from *a*; but to the south of this point the dead bugs were not so abundant, while, on the other hand the live insects were just as numerous. This was perhaps due to the fact that this was the highest portion of the field, and was considerably dryer than at *a* and points further north. The chinch-bugs had spread over about three acres, and had killed all the grass west of the wavy line passing through F and G. The grass was parched and brown and the victorious chinch-bug hosts were steadily advancing eastward.

A careful examination of the grass west of *a*, revealed the fact that the number of bugs dead with the fungus increased as the sorghum (E) was approached; that is, the number of fungus-covered bugs counted on a given area—one square foot— constantly increased until the point *b* was reached, at which place the greatest numbers were observed, and then gradually decreased, practically disappearing at *c*, as we passed into the corn (D). At *b* two hundred and seventy-one whitened bodies were counted on the surface of the ground scattered over a single square foot. The ground at this point, however, was thickly strewn with sorghum leaves and was quite damp.

Another interesting point observed at this time was the comparative age of the fungus along a line from *a* to *c*. At *a*, as already indicated, the fungous growth was fresh, and in some instances the mycelial threads were but just starting, being but barely visible on the bodies of the dead bugs. As we approached *b*, however, it became evident that the fungus on the insects was of a much older growth. At a point midway between *a* and *b*, bugs dead with the fungous disease were found, from which the spores were not easily detached; while at *b* the spores were easily shaken off, and in most cases the ground where a dead bug had been lying was so dusted with them as to present a whitish, mouldy appearance. From *b* to *c* the fungus was of a still earlier growth. Along a dead furrow at the latter point, several dirty whitish, mouldy spots on the surface of the ground under a fallen corn leaf or stalk, or other rubbish, was all the evidence that could be found that the fungus had been present.

Comparatively few chinch-bugs dead with the white fungus were found in corn B, C, H, I, and J, not more than half a dozen being taken from each field. Their whitened bodies were perhaps more abundant in corn H along the south side in the immediate vicinity of the meadow G, than in any of the other fields examined. Live chinch-bugs, mostly pupae and adults, were quite numerous throughout all these fields. All the meadows adjoining corn on this farm were more or less injured from the attacks of the chinch bug, but little or no fungus was found in any of them except those already mentioned, F and G. Only an occasional live insect was seen in the wheat stubble A, in which the infection had been placed, and in the stubble of oats and wheat south and east of the house.

A dozen or more insects dead with Sporotrichum were found in a corn field belonging to Mr. Wm. Kelley one mile east of the Heth farm, and several were taken from the surface of the ground under stools of foxtail-grass along the roadside in the immediate vicinity of the same field.

The Heth farm as well as the surrounding country was again examined by Mr. Johnson and myself October 10. The fungous attack had not increased in intensity, so far as could be ascertained, in the meadows F and G; but the chinch-bugs had continued their ravages, and the irregular line marking the boundary between the infested area and the remainder of F and G had moved several feet eastward. The whitened bodies of dead bugs could be easily found on the ground by parting the grass at any place west of the division line, and they were quite abundant under leaves and rubbish in the sorghum stubble E, but the growth in this latter place had the characteristic weathered appearance of over-ripeness. In corn D an occasional trace of the fungus could be seen, and several chinch-bugs imbedded in it were found under stools of grass along the roadside between C and D. In corn B the fungus was found in shocks and in the open field in the stubble, but that from the surface of the ground under the shocks was of a much more recent growth than that taken from the open field. An occasional bug dead with this disease occurred on the ground in the same field.

One mile east, on the Kelley farm, to which reference has been made, chinch-bugs enveloped in the white fungus were found in corn on the ground in low, damp places, under fallen leaves, weeds, grasses, and other rubbish. They were also quite abundant along the roadside under stools of grass, but on the whole the fungus was not so plentiful as twelve days previous, and the growth was comparatively old. In an adjoining wheat field, a considerable number of dead insects were found attached to the blades of young wheat plants, and on the ground between the drill rows. After counting both live and dead chinch-bugs on a given area, it was estimated that about twenty-eight per cent. were dead. Subsequent examination of collected specimens showed, however, that the destructive agent in this instance was the gray muscardine (*Entomophthora aphidis*); but a few dead chinch-bugs from this field, showing no external trace of a fungous growth, developed a profuse growth of the white muscardine when placed on damp sand. This fungus was also found on chinch-bugs in all corn fields examined at this time for a distance of four miles south of the Heth farm, but was nowhere as abundant as on the latter place.

It has been already reported on page 53 that the white fungus of the chinch-bug was very widely distributed throughout adjoining counties at this date, and the general tenor of our observations at this time supports the hypothesis that the Heth outbreak was a spontaneous one, arising under the influence of especially favorable conditions, which were substantially as follows: (1) an abundance of food—wheat, corn, sorghum, and grass, into which the chinch-bug hosts passed successively after each harvest, thus keeping them somewhat concentrated; (2) the appearance of the fungous disease on chinch-bugs along a dead furrow running north and south through *c*, at a time when the destruction to corn was about complete, and when myriads of insects were passing this point, going into sorghum, E; (3) the concentration of the bugs on this narrow strip of sorghum—practically accumulating the chinch-bugs of thirty-five acres on these nine rows; (4) the stripping and cutting of this sorghum at a time when the insects were most numerous, which knocked them to the ground in great masses, where they remained under and among the leaves for several days; (5) the occurrence of a heavy rain September 12, followed by a high temperature, and another heavy rain September 16; and (6) the close proximity of the meadows F and G, into which the bugs passed, and where they remained on the damp ground under the grass.

It thus seems quite probable that this fungous outbreak had little or no connection with the infection material distributed during the early part of the season, but that it was simply a local development, under exceptionally favorable conditions, of the diffuse general spread of chinch-bug disease throughout this whole country at this time.

The three following (Nos. 78–80) are successive farmers' experiments conducted in the counties of Cumberland, Bond, and Clay. Individual reports on these farmers' experiments were not called for, and the report here given for 78 and 79 is based upon a single visit made by Mr. Johnson early in October. No. 80 was visited twice.

No. 78. A farmer's experiment, the original material for which was derived from our experiment No. 54. It was sent from this office June 25 to Mr. Thomas B. Wilson, Sr., Greenup, Cumberland county, and used by him to start a contagion box, according to our circular of directions (see p. 28). This box was kept in operation until late in July, live bugs and fresh food being introduced every other day, and the fungus-covered insects removed each time the box was opened. Several hundred such specimens were distributed during the first two weeks of July at the bases of the stalks and behind the leaves of corn along one side of a field adjoining wheat, in places where the chinch-bugs were most abundant. The ground was very dry, and the experiment was a failure, no trace of the fungus appearing so far as could be ascertained by weekly examinations kept up until late in September. These fields were examined by Mr. Johnson October 9, but no trace of insect disease could then be found, although adults and pupae of the chinch-bug were still abundant in the corn. The first fifteen rows adjoining the wheat had been completely destroyed, only here and there a stalk still standing erect. On this same visit several bugs imbedded in Sporotrichum were taken from corn shocks about two miles from town, in another direction.

No. 79. A farmer's experiment, conducted by Mr. W. E. Jackson, of Greenville, Bond county, with material from No. 54. About a dozen chinch-bugs, received from this office, were placed in a box July 1, prepared according to our directions, with live insects and fresh food, both of which were renewed as necessary, a few whitened bugs being taken out of the box each time that it was opened, and distributed in the corn. July 15 about a hundred fungus-covered specimens were placed behind leaves and on the ground, where the bugs were most abundant, in a field of corn on Mr. A. H. Jackson's farm, a quarter of a mile south of his house. Two similar distributions were made in this field July 28 and August 12. The corn was examined every week, but no indication was found that the disease was spreading, the chinch-bug attack, in fact, increasing steadily week by week. The corn was cut and shocked about September 3, in order to save it for fodder, and the bugs collected in the shocks, where they remained until late in fall. This field was carefully examined by Mr. Johnson October 10. The corn had been thoroughly wet by heavy rains soon after cutting, and the ground was still very wet under the shocks. There was little or no grass in the open field, and consequently very few chinch-bugs. They were quite abundant in the shocks, and numerous in an adjoining meadow, but not sufficiently so to do any appreciable damage. A careful search was made in shocks all over the field, but no traces of the white fungus were found, except in four shocks in the northeast corner, and here the greater part of the diseased insects were found in the third shock of the first row on the north, counting from the east. Several hundred chinch-bugs and three beetles (*Coccinella 9-notata* Hbst., *Ataenius stercorator*, Fab., and *Epicauta vittata*, Fab.) completely enveloped in the white fungus were taken from the ground and from behind leaves in this shock. Half a dozen chinch-bugs dead with Sporotrichum were taken from three other shocks in the immediate vicinity. The disease was also found in small quantities in corn shocks directly east on the opposite side of the road; and three whitened bodies were taken from a low damp furrow or ditch in the meadow adjoining on the north. A few fungus-covered bugs were found in corn shocks in an orchard near the house, and quite a number were found in shocks and on the ground in low damp places in a corn field one mile east.

Just why the disease should have been so abundant in the single shock mentioned above, and totally absent or nearly so in all other shocks in the field, we will not attempt to explain. Suffice it to say that the infection was distributed along this side of the field, and that all the other shocks in the same row were similarly situated and cut at the same time.

The experiment was a failure, and did not arrest the ravages of the bugs in the least. A part of the fungus-covered bugs collected in these fields were placed in W. E. Jackson's hands for future experimental purposes.

No. 80. A farmer's experiment, made by Mr. C. M. Filson, Xenia, Clay county, the original material for which was derived from this office. The box was carefully prepared according to our circular of directions (see p. 28), and infected June 20 with about a dozen bugs from No. 54, this series. The box was kept on the damp floor of a cider house. Bugs enveloped in the white fungus were removed every fourth or fifth day, and live insects and fresh food were added as required This box was kept in active use until about September 1. It was examined by Mr. Johnson September 27, at which time, although the insects were all dead and the food was dried up, it was still in good condition, and about a hundred chinch-bugs, thickly imbedded in the muscardine fungus, were taken out and given to Mr. Filson for future use. Several hundred fungus-covered bugs from this box were distributed in corn along the north and south sides of a field on Mr. Filson's farm about July 1, and about ten other distributions were made at regular intervals after a rain or on damp mornings, with bugs from the same source. Chinch-bugs entered this corn from an adjoining wheat field, literally covering the stalks in many places. This field was examined about every third day, but no trace whatever of the white fungus was detected. As a consequence, this experiment was abandoned, and other measures were taken by the owner to arrest the ravages of the chinch-bug (see No. 89). Mr. Johnson examined this field September 27 and again October 6, but did not find a single infected insect. The corn had been cut at the time first mentioned, but the bugs were still quite abundant in the shocks, and foxtail and other grasses along the fences and in low damp spots throughout the field were thickly covered by them.

Several fungus-covered bugs were found by Mr. Johnson September 27 in grass along the roadside more than a mile from Mr. Filson's farm; and again on the Filson farm also during the latter part of November. December 10 Mr. Filson writes: "I could see no effect of the disease on my farm until after the recent rain. At the present time I find chinch-bugs covered with fungus in all the shocks in corn adjoining timber."

2. EXPERIMENTS WITH BARRIERS AND TRAPS.

The six following (Nos. 81–86) are field experiments with barriers and traps conducted by us this summer on the University farm. The appearance of the chinch-bug here, for the first time since 1883 in numbers sufficient to do noticeable injury, gave us an especially favorable opportunity for careful experimental work. Experiments 81, 82, and 83 were made on a small scale to test the efficiency of the furrow and post-hole method for the arrest and destruction of chinch-bugs while escaping from fields of small grain at harvest time. Numbers 84 and 85 were made to test the value of the coal-tar barrier, and 86 was a practical test of the two combined.

No. 81. This is a furrow experiment made July 10. A patch of wheat stubble ground (B, Plate II., in the vicinity of *c*), 4x6 feet, was cleared off with a spade so that the surface was hard and smooth. Around this we dug up and pulverized a narrow strip of ground, in which a dusty furrow was made three inches deep inside and six inches outside, enclosing the entire patch. The outer face of this furrow had a slope varying from 50° to 60°.

At 2:10 P. M. we released in this enclosure over a pint of chinch-bugs* collected from corn adjacent, and observed their operations in the ditch. Probably one-fourth of those collected were adults, the remainder being of various ages—mostly pupae, or

---

*The numbers of chinch-bugs used in these experiments were determined by counting those in a given measure, 10 c. c.; a pint being thus ascertained to contain about 132,500; a quart 265,000; and a bushel 8,480,000.

in the stage immediately preceding. The adults were much the more active, the immature forms tending to accumulate and pile up on each other in the ditch.

The greater part of the chinch-bugs presently deserted the interior of the enclosure and attempted to escape from the ditch, forming a continuous belt in the bottom one to three inches wide and, where thickest, two or three layers deep. In their efforts to escape, the adult bugs persistently climbed up the outer face of the furrow again and again, without cessation, falling back each time to the bottom as the dust gave way beneath them, the result being finally to accumulate a slope or talus of dirt at the bottom of the furrow of an incline sufficiently gradual to permit them to climb it easily. In this manner they slowly advanced upward, until in an hour and a half from the beginning of the experiment a few escaped at one corner by climbing up a kind of ladder-way of small clods and roots projecting from the surface. Not over fifteen or twenty thus released themselves, when the clods were undermined and fell, breaking the passageway.

An hour and three-quarters from the beginning, a post-hole was made in the furrow at one end of the enclosure. The chinch-bugs nearest it presently fell in, and as others advanced to take their places—apparently impelled by the pressure from their neighbors—they were also trapped. The impulse was thus gradually passed along the struggling line until within a few minutes there was a definite movement of the entire body of chinch-bugs for about three feet on each side of the hole towards and into it. By 4 o'clock, probably half of the chinch-bugs in the enclosure had been trapped. This movement had so greatly diminished the progress of those attempting to ascend the side of the furrow, that at this time they had nowhere generally advanced beyond an inch below the upper edge. Without the post-hole it is likely that they would have begun to make their escape in considerable numbers in about two hours from the time the experiment began.

We collected and brought to the laboratory from the post-hole trap and from the furrows about one pint of chinch-bugs, leaving the remainder in the enclosure. Next morning the greater part of these—probably all except the adults—were dead in the bottom of the furrow, killed by exposure to the sun

No. 82. July 11, two parallel furrows twenty-five feet long were made in a thoroughly pulverized strip of ground, in wheat stubble (B, Plate II.), by dragging an eight-inch log back and forth through the dirt. These furrows were connected at their ends by transverse furrows of the same character, thus enclosing a strip of solid, smooth ground between them a foot and a half across. The furrows were two inches deep inside and five inches outside, the outer slopes varying from 40° to 54° 30'.

Collected a quart and a gill of chinch-bugs (estimated number 300,000) from adjacent corn, C, and placed them on the strip enclosed by the furrows, distributing them the whole length of the plat. A few adults flew at once, and several others made the attempt, as, indeed, adults had occasionally done the day before. In twenty minutes probably nine-tenths of them were dead upon the ground, evidently from the heat of the sun. Most of them had died on the hard earth between the furrows without reaching the latter. Two-thirds of those in the furrows at this time were adults. Those dead from the heat were nearly all young, but an occasional adult was seen among them.

The principal movement of the imprisoned chinch-bugs was at first to the north, in the furrow on that side, but presently they abandoned their attempt to scale this slope, and all remaining alive resorted to the south furrow, collecting chiefly at two points. This was evidently due to the greater heat of the north furrow on account of its more direct exposure to the sun. As the chinch-bugs work at a somewhat steep slope they gradually undermine it, leaving an overhanging ledge which they cannot scale, but which they gradually work down in their efforts to climb the bank.

The temperature of the earth determined by simply laying a thermometer on it in the sun, was 116° Fah. at 11:05. If the thermometer were barely buried in the dust it was 122°. The air temperature at the same time was 91°. Sky cloudless, and a gentle wind.

To verify the effect of the hot earth and sun upon chinch-bugs, an additional pint was collected and put into the small enclosure used the preceding day. These were killed as above, and almost as rapidly. Wherever a rootlet, or any other solid substance, projected above the surface, it was thickly covered with chinch-bugs, and a stick of any kind thrust in among them would be immediately blackened by them, as they crawled upward, collecting at the top, and dropping off as they crowded each other outward. Taking advantage of this fact, a small trap was arranged by inclining sticks over a dipper of coal-tar. Two or three fluid ounces of chinch-bugs were collected in this way in the course of half an hour. This is, however, a very much less rapid method than the post-hole trap.

We estimated that at the rate of action of the chinch-bugs in these furrows a single man could certainly supervise eighty rods along the edge of a field, and probably twice as great a distance.

No. 83. July 12 a quantity of chinch-bugs was placed in the large enclosure described under experiment 82, the first at 6:25 A. M. and a second lot at 7, and a coffee can was sunk in the furrow at one end of the plat.

At 7:40 chinch-bugs placed in the southern furrow and a belt of coal-tar poured along the middle of the enclosed space. Day clear and rather windy. At one place where the slope was 58°, chinch-bugs began to escape almost at once; but by steepening the furrow at this point with a boe, we confined them permanently, only a very few escaping by making here and there a temporary passageway which permitted now and then one, at long intervals, to emerge. Such passageways were presently undermined, and the number escaping was entirely insignificant.

Chinch-bugs began to collect in the can as soon as they were placed in the furrow, but the angle of the furrow near it arresting them, a second one was placed in the middle of the furrow. Into this can they fell in quantity, presently marching towards it from the right and left, thinning out the crowd in the furrow for a distance of nine feet on one side and ten to twelve feet on the other, until by 9 o'clock the attack was practically broken all along the line by the capture of nearly all of the bugs.

At 6:45 the temperature of the ground—the thermometer being lightly covered with dust—was 79°; that of the air, in the sun, 82°.

At 8:15 the temperature of the dirt, taken as above, was 97°, and that of the air, 85°—thermometer erect in the sun.

At 9, chinch-bugs had begun to die where most exposed to sun. Dirt was here 108°; air 87½°, with thermometer erect.

No. 84. July 12, a patch of wheat stubble was cleared off, as in experiment 81, and at 7:30 P. M. a belt of coal-tar two inches wide was put down, forming an oval, enclosing a space twenty-five feet long by two feet across. Post-holes about ten inches deep were dug at either end, with a common post-hole digger. Coffee cans, about six inches in diameter and seven or eight inches in depth, were placed in the post holes so that the entrapped chinch-bugs could be easily removed and measured.

At 7:45 P. M. half a pint of chinch-bugs was distributed over the hard, smooth surface within the enclosure. At first they went in all directions, and many ran headlong into the tar and were destroyed; but the greater number were more deliberate, and moved up and down the tar line without making any attempt to cross it. By 7:50 a large proportion had passed from the middle of the enclosure to the edge of the tar, principally on the south side, but forming a belt, as it were, around the entire enclosure, the general movement being eastward.

At 8 o'clock they were less active and were most abundant at the ends of the oval, but very few had fallen into the post-holes. There was no disposition to climb rootlets or other projections above the surface.

At 8:30 they were still less active, and were collecting together in masses on small lumps of earth and in depressions on the surface. No general movement observed at this time. About fifteen hundred chinch-bugs were removed from each can. They were most abundant in the east and west ends and along the south side of the oval. The slightest disturbance, such as the movement of a finger on the ground in their midst, caused the greatest confusion among them.

The sky was clear with a gentle breeze from the west. Temperature of air at 7:30 was 82°, the thermometer held erect; surface 82°, as determined by simply laying the thermometer on the ground.

At 8:30 both air and surface were 74°, observations being taken as above.

At 5 A. M., July 13, the young were more active than the adults, and a great many bugs were still collected on lumps of earth and in depressions on the surface. A few were crawling and falling into the post-holes, about as many having been trapped in eight hours and a half during the night as had been caught the previous evening in three-quarters of an hour.

At 6 o'clock the temperature of the air was 74°, thermometer erect and about four feet above the surface. The soil temperature was 78°, thermometer slightly buried. The chinch-bugs were now very much more active and were moving in considerable numbers toward the east end of the oval, about six times as many (9 000 in round numbers) having collected in the post-hole at this point since 5 o'clock as had been entrapped during the entire night. Their activity steadily increased, and in a few moments there was a regular procession fourteen feet long moving to the eastward along the tar line towards one end of the oval, and to the westward for a distance of six feet towards the other, leaving an intermediate space of about five feet where there was no appreciable tendency in either direction Very few bugs passed the tar line, although it was dry and could easily have been crossed. It served, practically, as an impassable barrier.

At 9 A. M. the temperature of the air was 85°; surface 112°. The oval was almost entirely freed from chinch-bugs, the most of them having fallen into the post-holes. Two-thirds of the entire lot that had been placed in this enclosure the previous evening, were taken from the cans, the great majority having fallen in since 6 A. M. If the three thousand chinch-bugs taken from the post-holes at 8:30 P. M. the preceding day and those still remaining within the oval are taken into consideration, it is clear that only a very small number escaped.

No. 85. July 13, an experiment similar to No. 84 was made on an oval half as large, with post-hole in one end. Slight rain at 2 P. M., just enough to settle the dust. Sky cloudy, with light breeze from southwest; temperature of air 90°; surface 87°.

At 2:30 P. M. renewed barrier by pouring coal-tar over line used the day before, for a distance of twelve feet on either side, and across the ends; but at the east end the tar was poured over the ground, as no line had been previously made at this point. One gill of chinch-bugs was distributed on the surface of the hard ground enclosed. They were very active, and in ten minutes the center of the oval was comparatively free, the insects forming a band next the tar line around the entire enclosure. By 3 P. M. the bugs had mostly collected on the south side and in the ends, being most abundant in the east end, where they were very active. The general movement at this time was to the eastward, but many were tumbling into the can at the west end . A strong wind from the southwest blew many insects over the line.

Outside the tar line myriads of young bugs just from the egg were moving southward in the direction of the adjoining corn C (Pl. II.), literally covering the ground in many places. Not a single insect attempted to cross the tar, although in their confusion they scrambled about in all other directions; but where no barrier intervened they passed rapidly along towards the corn.

Two-thirds of the bugs within the oval had collected at either end by 3:30 P. M., and about two thousand had fallen into the can.

At 4:30 P. M. the barrier was in good condition except at one place where the coal-tar had been poured over the loose ground and was now getting quite dry. A few insects attempted to cross the line, but either retreated or went pell-mell into it and were destroyed. One-third of the entire lot had collected in the east end, and the others were scattered about the oval. About five thousand were taken from the can and the experiment was left over night.

July 14, at 9:30 A. M., of the twenty-one thousand bugs left in the oval the pre vious night, less than two thousand remained, one-half of these being in the can. The remainder had escaped during the night and early morning through a passageway at the east end, where the tar had become dry and where the wind had blown fine particles of dirt over the surface, completely covering it.

At 10 A. M., sky clear and a gentle westerly breeze. Temperature of air 84°; surface 106°; soil 117°.

No. 86. July 10, a strip of ground between the spring wheat B and corn C (Plate II.), three feet wide, was thoroughly and deeply pulverized by means of a harrow-toothed cultivator drawn by one horse and a twelve-foot plank drawn endwise, the driver riding the harrow or plank when necessary. Next, a log about six feet long and eight inches through was dragged endwise back and forth in this strip, the driver riding it, until a deep furrow had been made. The sides of the furrow were then dressed up here and there with a hoe.

Similar furrows were made in the fifth and sixth rows, and a narrow line of coal-tar was poured along the bottom of the furrow in the latter row, from an ordinary two-gallon sprinkler without the nozzle. On the first application one gallon of tar was sufficient for a line ten rods long, and thereafter for about twenty rods. The tar very soon formed a crust, but remained in good condition and completely checked the advance of the chinch-bugs for twenty-four hours or longer.

Holes about a foot deep were made in each furrow with an ordinary post-hole digger at intervals varying from ten to twenty feet, according to the abundance of the bugs.

A strip of winter wheat (A) of about four and a half acres, badly infested with chinch-bugs, was cut June 27 to July 3. The bugs then attacked the narrow strip of spring wheat (B)—about one rod wide, running the entire length of the field,—which they completely ruined. This was cut July 7 and burned over the following day. Many bugs were destroyed, but the great majority of them moved into the corn C, blackening the stalks in the first two or three rows.

The furrow beside the first row did not check their advance to a very great degree, from the fact that it had been defaced and broken down to some extent, and was strewn with straw and other rubbish from the wheat. The furrow beside the fifth row and tar line by the sixth, however, completely arrested their advance, and practically kept them confined to the first five rows. A quart or more could have been easily collected in a few minutes by jarring the stalks and catching the bugs in a pan.

The insects worked away in the furrow, endeavoring to escape in much the same manner as described under experiments 81 and 82. An occasional one made good its escape by means of a projecting rootlet, or the rubbish strewn about, but was repelled by the tar line in the next row, which seemed to be regarded as an impassable obstacle. In both cases there was a general movement up and down the lines, and the bugs were constantly falling into the post-holes, a pint or more being entrapped in each, where they were killed with a strong mixture of kerosene and water or by a little coal-tar poured upon them.

In the furrows, where the bugs were directly exposed to the sun, a great many were killed by the extreme heat, the tender larvae succumbing first, but even adults dying finally.

These furrows were dressed up here and there from day to day with a hoe, as was necessary, and the tar line was renewed about every twenty-four hours. A slight rain fell July 13, just enough to lay the dust, and the furrow in the fifth row no longer restrained the marching horde. The ground was literally covered with young bugs either in the pupa stage or the moult just preceding, and their advance was southward toward the center of the field. The tar line, however, remained unaffected, and proved the same impassable barrier to the advancing hosts as when first put down. A state of utter confusion prevailed, and the bugs ran restlessly up and down the tar front, tumbling into the post-holes, where they were finally destroyed, or being speedily killed in the furrows by the excessive heat, as they ran here and there over the ground.

The insects made good their advance in the eastern half of the field, where no barriers obstructed their course, and completely covered the corn as far as the ninth and tenth rows inward. The average yield in such places was reported at the end of the season by the farm superintendent as about twenty per cent. less than that of corresponding rows in the upper part of the field, where the barriers had been used. The chinch-bugs, on the other hand, were originally far less numerous in this part of the wheat adjacent than at that end of the field where they were destroyed as described above.

The five following (Nos. 87-91) are successive farmers' barrier experiments made to arrest the advance of chinch-bugs as they moved from wheat to corn in late June and early July.

No. 87. Made by Mr. Samuel Bartley, of Edgewood (see Nos. 63-67), June 27. A narrow strip of ground in corn along the side adjoining wheat (No. 63), was deeply pulverized, and through this a deep furrow was afterwards made by dragging a log endwise. The sides of the furrow were as steep as they could be made without caving in. The wheat was cut the following day (June 28), and the chinch-bug hosts started for the corn field. Their advance was completely checked for a time, and they accumulated in great numbers in the furrow.

Mr. Marten visited this field June 29. The furrow was then in fair condition, and contained myriads of bugs endeavoring to escape. There were many insects in the wheat stubble and in a narrow strip of grass between the wheat and corn, but comparatively few on the latter crop.

No provision had been made for the destruction of the bugs in the furrow, and a slight rain June 30 breaking down the sides, in a short time the traveling horde made good its escape, almost completely destroying the corn as it advanced.

No. 88. This is a furrow experiment, made by Mr. James Smith, of Farina (see Nos. 59 and 60). About June 28, just before wheat harvest, Mr. Smith abandoned his contagion box (No. 59) and plowed furrows between wheat and corn, wheat and oats, and oats and corn. The furrows were made about eight inches deep, with a shovel plow. A log drawn by one horse was dragged back and forth through these furrows for about a week. Myriads of bugs were crushed, and many died from exposure to the heat, as they were confined in the furrows. Very few bugs crossed these ditches into the corn, and less than a quarter of an acre was injured by their attacks. A slight shower had fallen soon after the log was started, stopping operations for a short time, and a considerable number of chinch-bugs then passed the furrow, but the ditches were opened again as soon as possible, and the dragging was resumed.

At the time of Mr. Marten's visit, July 11, the log had not been used for several days, and the bugs were crossing the ditches in great numbers and were accumulating on the corn, where they did considerable damage later in the season.

No. 89. This is a barrier experiment conducted by Mr. C. M. Filson of Xenia, in corn (see No. 80) adjoining wheat which was cut June 15. The chinch-bug horde came into the corn immediately and ruined ten or twelve rows. Mr. Filson thoroughly pulverized a narrow strip of ground in the twelfth row on the south side of the field, and along the outer edges of the east and west sides, and through this a log was dragged until a deep furrow was made the entire length of the three sides. Post-holes were then dug in the furrows about ten feet apart, into which the traveling bugs fell in great numbers, where they were destroyed with kerosene emulsion or by crushing. While the furrows were in process of construction, quite a number of bugs succeeded in crossing the ditch and accumulated on the first two or three rows beyond. These were destroyed with kerosene emulsion, applied by means of a brush-like broom, made of prairie grass, dipped into a pail containing the emulsion and shaken over the bugs on each hill. Many insects fell on the ground during this operation and were killed by the emulsion. The furrows were kept in good condition for two weeks. The progress of the incoming horde was practically arrested, and very few bugs were seen in the field the latter part of the season. The corn yielded about twenty bushels to the acre, which was more than the average for that neighborhood.

A field of corn, adjoining this same wheat field on the south, in which no measures were taken to arrest and destroy the chinch-bugs as they came from the wheat, was ruined, excepting only a small part which was thought worth cutting for fodder.

No. 90. Mr. H. H. Mayo, of Falmouth, in Jasper county, made a deep furrow in a well-pulverized strip of ground in corn. Wheat adjoining was cut about June 23, and the bugs entered the corn in great numbers. The furrow completely checked their advance for a time, and myriads of young were seen dead in the furrow from exposure to the extreme heat. A slight rain fell shortly after the wheat was cut, after which the furrow was not reconstructed, and the pests had free passageway into the corn. Over four acres were completely destroyed in a few days, and the attack spread throughout the twenty acres, from which less than half a crop was taken.

No. 91. This is a barrier experiment made by Mr. Thos. B. Wilson, Sr., of Greenup (see No. 79), in corn adjoining wheat. The wheat was cut June 25, and the chinch-bugs made rapid advance into the corn. The ground between every fifth row from the edge, for a distance of twenty rods, was thoroughly pulverized and deeply furrowed June 26. The bugs collected in these furrows in great numbers and were killed by dragging a log back and forth. This was kept up for eight days. The first fifteen rows were entirely destroyed, owing to the fact that the insects accumulated here before the furrows were made. The corn in the remainder of the field was far better than the average in the county, and yielded fifty bushels to the acre.

In concluding this report, I desire to express my personal obligation and that of the Station to Mr. John Marten, by whom, as has doubtless been noticed, by far the greater part of the detailed experimental work was done; and to Mr. W. G. Johnson, by whom the notes of experiments were converted into the first draft of the detailed descriptions.

S. A. FORBES, PH. D., *Consulting Entomologist.*

EXPLANATION OF PLATES.

PLATE I.

Farm of G. C. Wells, near Farina, Fayette county; site of Experiments Nos. 55-58.

PLATE II.

Portion of University Agricultural Experiment Station farm, at Urbana, Illinois, showing the site of Experiments Nos. 81-86.

PLATE III.

Farm of G. W. Heth, near Edgewood, Effingham county; site of Experiment No. 77.

PLATE IV.

Farms of Silas Hurd, and Frank H. Robinson, near Odin, Marion county; site of Experiment No. 75, and of muscardine outbreak No. 76.

PLATE V.

Fig. 1.   Mason fruit-jar with altered cap.
Fig. 2.   Chinch-bug imbedded in Sporotrichum bearing heads of ripe spores.
Fig. 3.   A fragment of Sporotrichum from chinch-bug, spread out to show structure.   Highly magnified.

PLATE VI.

Culture mass of corn-meal batter from fruit jar, covered with Sporotrichum, showing its mode of growth on solid media.

PLATE VII.

Isaria Forms of *Sporotrichum globuliferum*, Speg.
Fig. 1 and 2.   On buried pupae of the apple leaf skeletonizer (*Canarsia hammondi*, Riley), as it appeared on the surface of the ground.
Fig. 3.   On pupa of same, the fungus bursting through the cocoon.
Fig. 4.   On buried pupa of same, uncovered and showing the fungus growth as it appeared below the surface.
Fig. 5 and 6.   On June beetles (Lachnosterna) found in ground.
Fig. 7.   On pupa of walnut caterpillar (Datana).   From laboratory infection experiment.
Fig. 8.   From white grub dead under ground in breeding cage.   From laboratory infection.   The fungus immature and the spores not yet fully formed.
Fig. 9.   The same as Fig. 8, but in a later stage, the fungus being mature, and the spores ripe.

PLATE VIII.

Map of Illinois, showing area and extent of distribution of muscardine fungus in 1894.   The figures for each county indicate the number of townships to which such infection material was distributed in June and July.

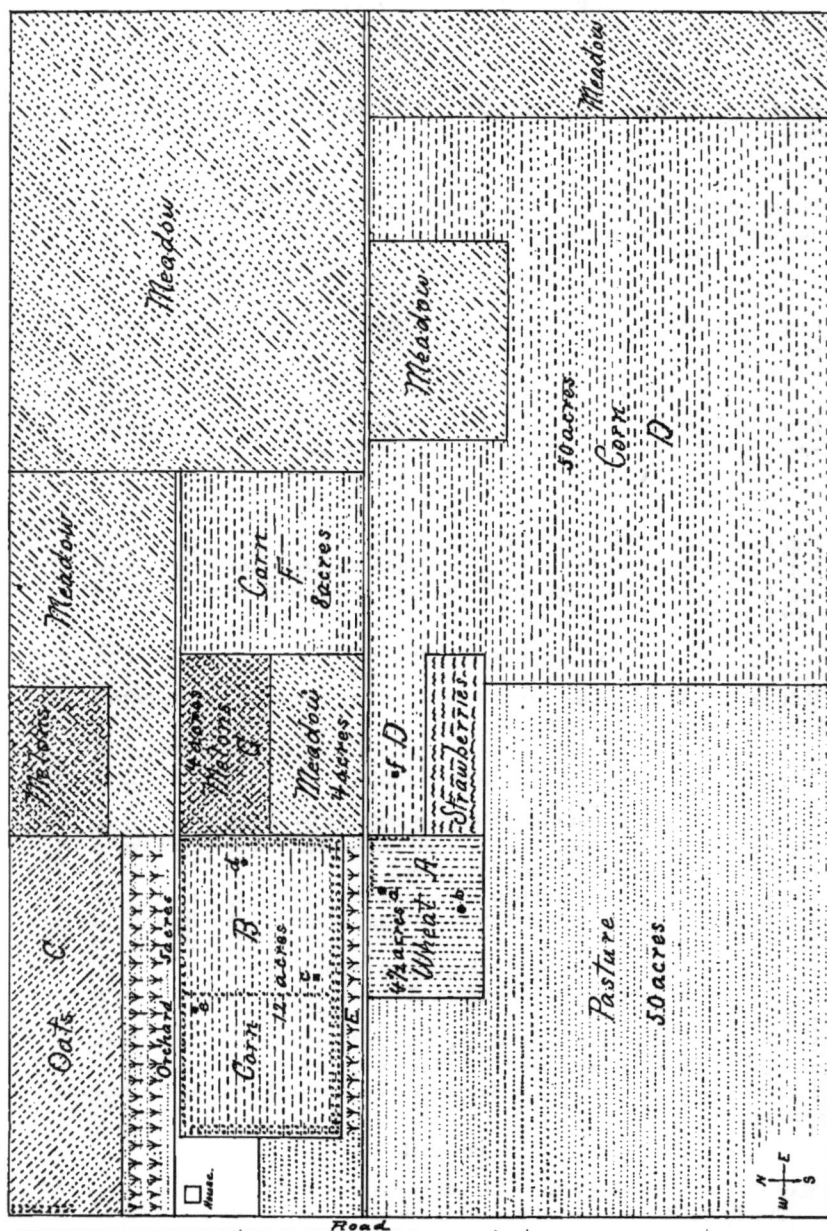

PLATE I. WELLS FARM.

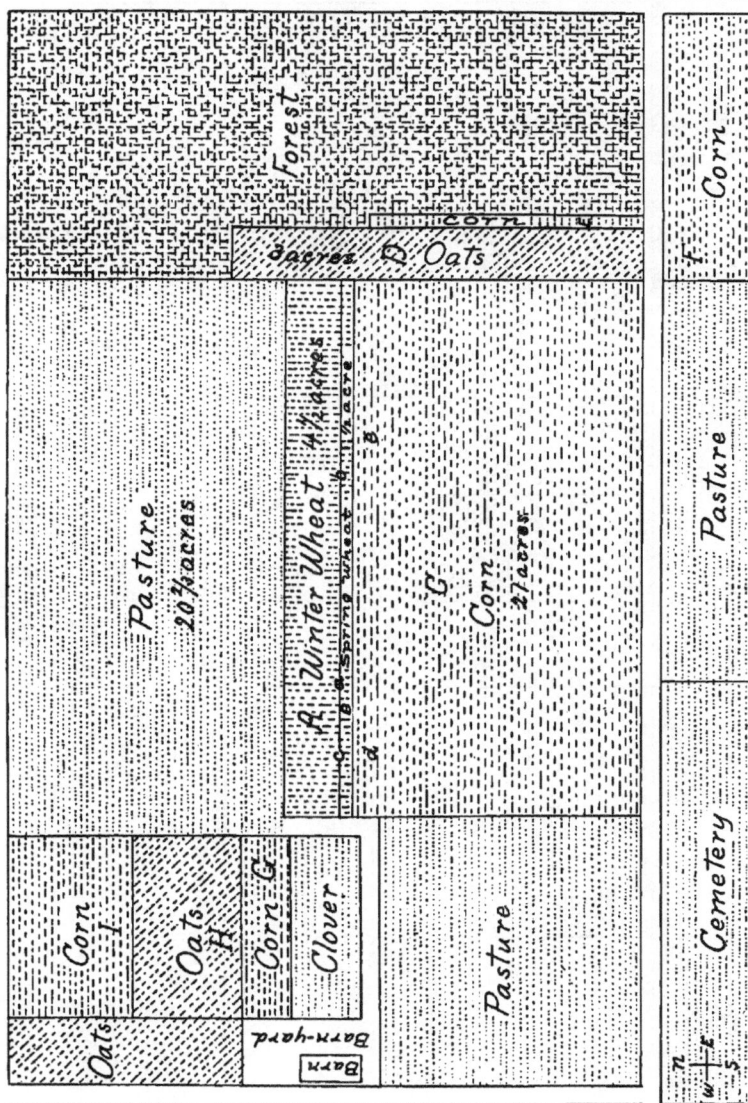

PLATE II.   EXPERIMENT STATION FARM.

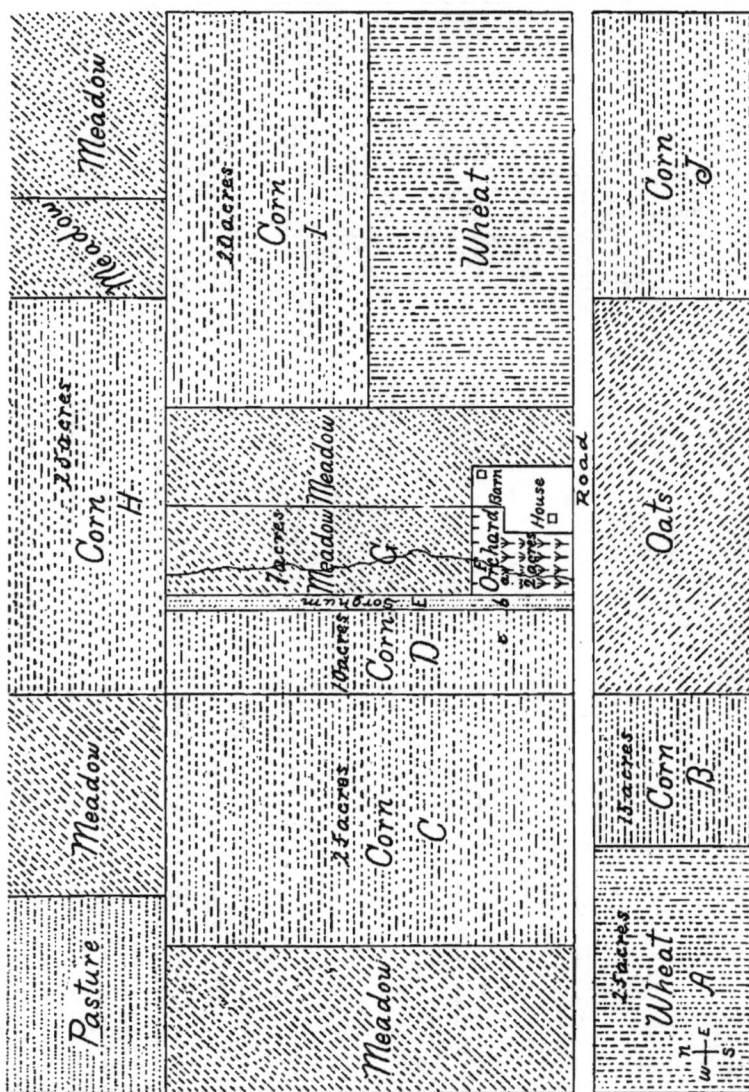

PLATE III. HETH FARM.

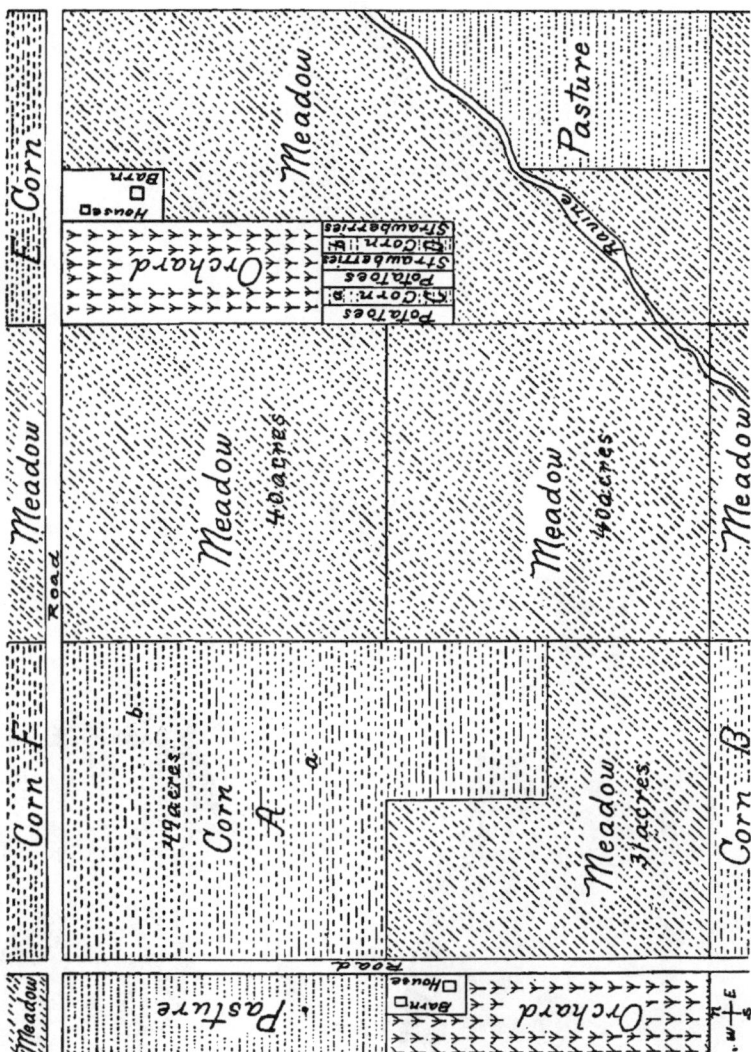

PLATE IV. HURD AND ROBINSON FARMS.

3

1   2

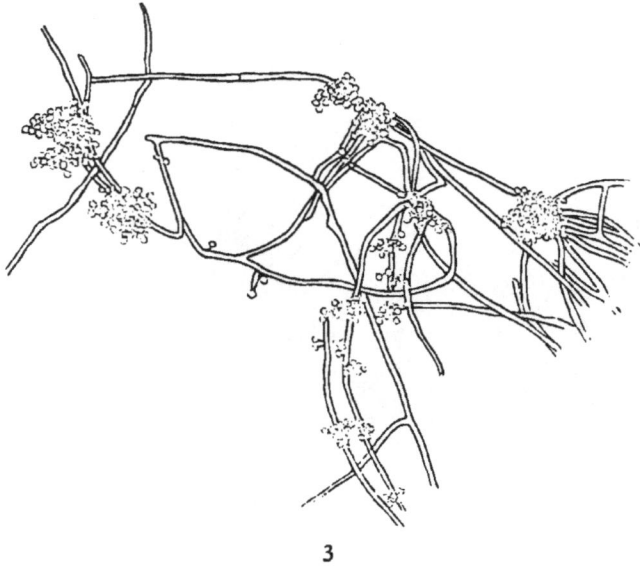

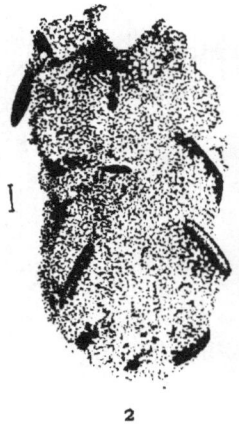

PLATE V.   SPOROTRICHUM AND CULTURE JAR.

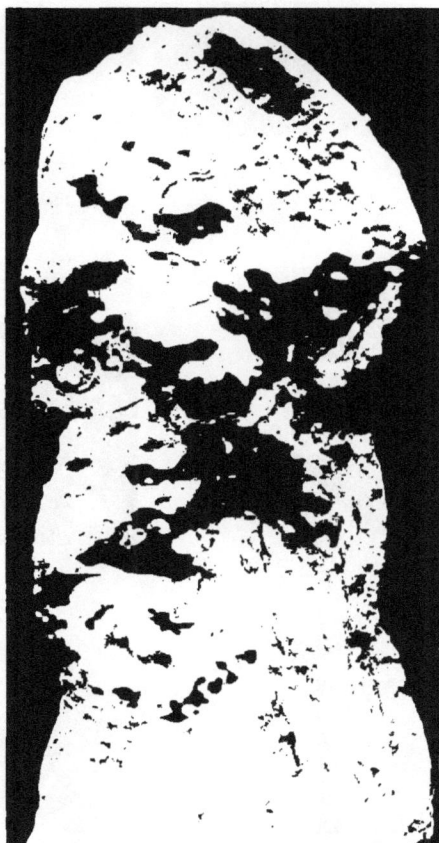

PLATE VI. CORN-MEAL CULTURE OF SPOROTRICHUM.

7

1

2

3

4 L.M.H.

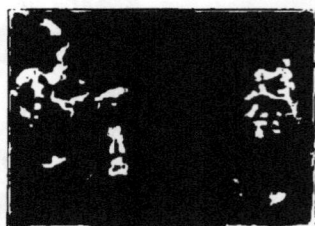

5   6

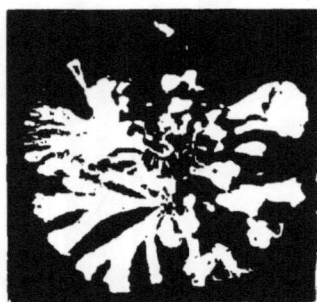

8

9

PLATE VII. ISARIA FORMS OF SPOROTRICHUM.

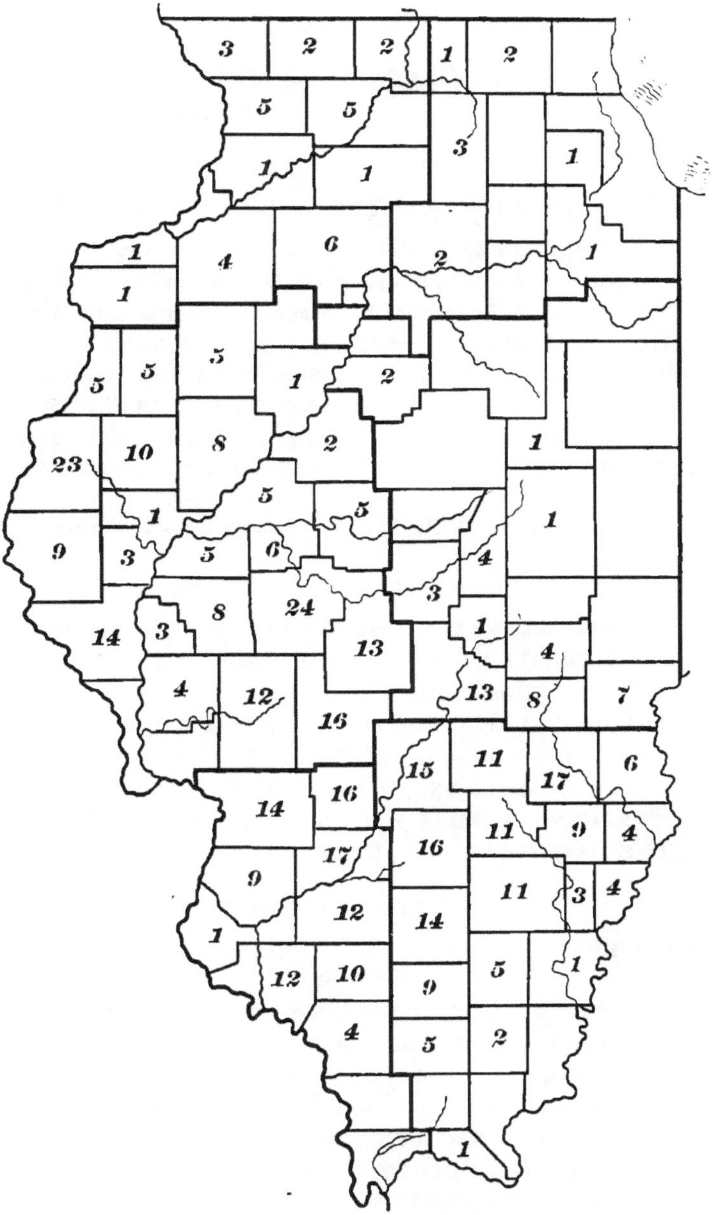

PLATE VIII.   MAP SHOWING NUMBER OF TOWNSHIPS IN EACH COUNTY
TO WHICH MUSCARDINE FUNGUS WAS SENT IN JUNE AND JULY.

ORGANIZATION.

BOARD OF TRUSTEES OF THE UNIVERSITY OF ILLINOIS.

NELSON W. GRAHAM, Carbondale, President.

JOHN P. ALTGELD, Springfield, Governor of Illinois.

JAMES W. JUDY, Springfield, President State Board of Agriculture.

SAMUEL M. INGLIS, Springfield, Superintendent Public Instruction.

RICHARD P. MORGAN, Dwight.   ISAAC S. RAYMOND, Sidney.

DR. JULIA H. SMITH, Chicago.   SAMUEL A. BULLARD, Springfield,

NAPOLEON B. MORRISON, Odin.   ALEXANDER McLEAN, Macomb.

JAMES E. ARMSTRONG, Chicago.   MRS. LUCY L. FLOWER, Chicago.

ANDREW S. DRAPER, LL.D., President of the University.

BOARD OF DIRECTION OF THE EXPERIMENT STATION.

THOMAS J. BURRILL, Ph.D., Urbana, Prof. of Botany and Horticulture, Pres.

E. E. CHESTER, Champaign, of State Board of Agriculture.

E. A. RIEHL, Alton, of State Horticultural Society.

H. B. GURLER, DeKalb, of State Dairymen's Association.

N. B. MORRISON, Odin, Trustee of the University.

ISAAC S. RAYMOND, Sidney, Trustee of the University.

STEPHEN A. FORBES, Ph.D., Urbana, Professor of Zoölogy.

EUGENE DAVENPORT, M. S., Urbana, Professor of Animal Husbandry.

THE STATION STAFF.

THOMAS J. BURRILL, Ph.D., Horticulturist and Botanist, President Board of Direction.

WILLIAM L. PILLSBURY, A.M., Urbana, Secretary.

EUGENE DAVENPORT, M. S., Urbana, Agriculturist.

STEPHEN A. FORBES, Ph.D., Consulting Entomologist.

DONALD McINTOSH, V.S., Consulting Veterinarian.

GEORGE W. McCLUER, M.S., Assistant Horticulturist.

GEORGE P. CLINTON, M.S., Assistant Botanist.

FRANK D. GARDNER, B.S., Assistant Agriculturist.

WILL A. POWERS, B.S., Assistant Chemist.

www.ingramcontent.com/pod-product-compliance
Lightning Source LLC
Chambersburg PA
CBHW021527090426
42739CB00007B/824